DATE			

CELL CYCLE REGULATION

This is a volume in
CELL BIOLOGY
A series of monographs
Editors: *D. E. Buetow, I. L. Cameron, and G. M. Padilla*

A complete list of the books in this series appears at the end of the volume.

CELL CYCLE REGULATION

Edited by

James R. Jeter, Jr.
Department of Anatomy
Tulane Medical Center
New Orleans, Louisiana

Ivan L. Cameron
Department of Anatomy
University of Texas Health Science Center at San Antonio
San Antonio, Texas

George M. Padilla
Department of Physiology and Pharmacology
Duke University Medical Center
Durham, North Carolina

Arthur M. Zimmerman
Department of Zoology
Ramsay Wright Zoological Laboratories
Toronto, Ontario, Canada

ACADEMIC PRESS, INC. New York San Francisco London 1978
A Subsidiary of Harcourt Brace Jovanovich, Publishers

ACADEMIC PRESS, INC.
111 Fifth Avenue, New York, New York 10003

United Kingdom Edition published by
ACADEMIC PRESS, INC. (LONDON) LTD.
24/28 Oval Road, London NW1 7DX

Library of Congress Cataloging in Publication Data

Main entry under title:

Cell cycle regulation.

(Cell biology monograph series)
Includes bibliographical references.
1. Cellular control mechanisms. 2. Cell cycle. I. Jeter, James R., Jr. II. Series.
QH604.C44 574.8'762 77-82415
ISBN 0-12-384650-1

PRINTED IN THE UNITED STATES OF AMERICA

Contents

List of Contributors

Numbers in parentheses indicate the pages on which the authors' contributions begin.

S. S. Barham (37), University of California, Los Alamos Scientific Laboratory, Los Alamos, New Mexico

D. Mona Baumgartel (167), Biology Department, Muir College, University of California, San Diego, La Jolla, California

I. L. Cameron (1), Department of Anatomy, University of Texas Health Science Center at San Antonio, San Antonio, Texas

J. A. D'Anna (37), University of California, Los Alamos Scientific Laboratory, Los Alamos, New Mexico

L. L. Deaven (37), University of California, Los Alamos Scientific Laboratory, Los Alamos, New Mexico

Donald F. Gerson (105), Faculty of Engineering Science, The University of Western Ontario, London, Ontario, Canada

Norman T. Gridgeman (203), Division of Biological Sciences, National Research Council of Canada, Ottawa, Ontario, Canada

Richard M. Gronostajski (185), Department of Biochemistry and Nutrition, Virginia Polytechnic Institute and State University, Blacksburg, Virginia

L. R. Gurley (37), University of California, Los Alamos Scientific Laboratory, Los Alamos, New Mexico

Patricia Harris (75), Department of Biology, University of Oregon, Eugene, Oregon

C. E. HILDEBRAND (37), University of California, Los Alamos Scientific Laboratory, Los Alamos, New Mexico

CHARLES W. HOFFMAN* (221), Department of Reproductive Biology, Case Western Reserve University, Cleveland, Ohio

P. G. HOHMANN† (37), University of California, Los Alamos Scientific Laboratory, Los Alamos, New Mexico

STEPHEN H. HOWELL (167), Biology Department, Muir College, University of California at San Diego, La Jolla, California

DANIEL W. ISRAEL (185), Department of Biochemistry and Nutrition, Virginia Polytechnic Institute and State University, Blacksburg, Virginia

ALLEN P. JAMES (203), Division of Biological Sciences, National Research Council of Canada, Ottawa, Ontario, Canada

J. R. JETER, JR. (1), Department of Anatomy, Tulane Medical Center, New Orleans, Louisiana

BYRON F. JOHNSON (203), Division of Biological Sciences, National Research Council of Canada, Ottawa, Ontario, Canada

MARGARIDA O. KRAUSE (61), Department of Biology, University of New Brunswick, Fredericton, New Brunswick, Canada

ENG-HONG LEE (203), Department of Pathology, University of Guelph, Guelph, Ontario, Canada

C. V. LUSENA (203), Division of Biological Sciences, National Research Council of Canada, Ottawa, Ontario, Canada

KENNETH S. MCCARTY (9), Department of Biochemistry, Duke University Medical Center, Durham, North Carolina

* Present address: School of Dental Medicine, University of Pittsburgh, Pittsburgh, Pennsylvania.

† Present address: Department of Experimental Biology, Roswell Park Memorial Institute, New York State Department of Health, Buffalo, New York.

KENNETH S. MCCARTY, JR., (9), Departments of Medicine and Pathology, Duke University Medical Center, Durham, North Carolina

G. M. PADILLA (1), Department of Physiology and Pharmacology, Duke University Medical Center, Durham, North Carolina

POTU N. RAO (133), Department of Developmental Therapeutics, The University of Texas System Cancer Center, M. D. Anderson Hospital and Tumor Institute, Houston, Texas

HELMUT W. SAUER (149), Zoologisches Institut der Universität, Würzburg, West Germany

ROBERT R. SCHMIDT (185), Department of Biochemistry and Nutrition, Virginia Polytechnic Institute and State University, Blacksburg, Virginia

PRASAD S. SUNKARA (133), Department of Developmental Therapeutics, The University of Texas System Cancer Center, M. D. Anderson Hospital and Tumor Institute, Houston, Texas

R. A. TOBEY (37), University of California, Los Alamos Scientific Laboratory, Los Alamos, New Mexico

R. A. WALTERS (37), University of California, Los Alamos Scientific Laboratory, Los Alamos, New Mexico

ANTHONY T. YEUNG (185), Department of Biochemistry and Nutrition, Virginia Polytechnic Institute and State University, Blacksburg, Virginia

A. M. ZIMMERMAN (1), Department of Zoology, Ramsay Wright Zoological Laboratories, University of Toronto, Toronto, Ontario, Canada

Preface

Regulation of the cell cycle is a complex interaction among the nuclear genome, the cytoplasmic pools, the organelles, the cell surface, and the extracellular environment. How these various units interact to govern the cell cycle is the topic of the chapters in this book. Obviously cell cycle regulation is too broad a subject to be covered in one volume. However, each of the contributors has summarized the recent work in their area of interest and has added some new work of their own. From this they have drawn conclusions and proposed hypotheses which should contribute to the further understanding of the control of the cell cycle.

Because of the basic nature of the information presented in this book, it should appeal to a wide variety of biologists and life scientists. In particular it should prove of interest to molecular, cellular, and to developmental biologists.

J. R. Jeter, Jr.
I. L. Cameron
G. M. Padilla
A. M. Zimmerman

CELL CYCLE REGULATION

1

Cell Cycle Regulation: Editors' Introduction to the Presentations

I. L. Cameron, J. R. Jeter, Jr., G. M. Padilla, and A. M. Zimmerman

Cell cycle regulation is perhaps too broad a subject for one volume. Yet the value of assembling contributions in a single book rests mainly on the ideas and concepts presented by each contributor. With this goal in mind, the authors were asked to briefly summarize the literature and salient features of their own field of research and to formulate conclusions and testable hypotheses as a basis for further understanding the regulation of cell cycle events. Elements of these presentations will eventually become material for a textbook, but, at present, the editors have sought to bring forth a source of provocative new information and ideas from within the individual chapters. We are fully conscious that our knowledge of cell cycle regulation is in a state of flux—a healthy state of affairs that we hope will continue for many years and to which this volume will become a notable reference point for continued investigations.

Much of the present work concerning cell cycle regulation centers around modulation of nuclear chromatin. In this vein, several chapters in this treatise deal with some aspect of chromatin modification and its effects on gene expression. One of the most dramatic changes in gene expression during the cell cycle is the drastic decrease in gene transcription as the chromosomes condense during mitosis. It, therefore, seems fitting that three chapters deal with the role of histones in chromatin structure and with postsynthetic modifications of histone residues (i.e., phosphorylation, acetylation, and methylation). It appears that such modifications lead, not only to structural changes such as those of chromosome condensation, but also to the regulation of gene expression.

McCarty and McCarty (Ch. 2) give an excellent and up-to-date review of the macromolecular structure of chromatin subunits called variously: "nucleosomes," "nu bodies," or "a chain of spherical particles." The nucleosome denotes a unit in which four classes of histones, consisting of H2A, H2B, H3, and H4 (the H1 histone is not involved in this assembly), provide a hydrophobic domain around which is wrapped or coiled a double-stranded helical native DNA molecule. The nucleosomes are ~8.0 nm in diameter and are linked together with a 1.5–2.0 nm diameter fiber. The nucleosomes are visualized when chromatin fibers are spilled out of lysed nuclei and can also be reconstituted by the combination of appropriate mixtures of native DNA and the four histones mentioned above. How is such a structure related to the 20-nm diameter fiber [termed the native chromatin fiber by Ris (7)] elegantly displayed in thin sections of cells prepared for electron microscopy? To help answer this question, Finch and Klug (5) indicated that the nucleosomes are condensed into a supercoil or "solenoid" structure of 20-nm diameter under appropriate conditions.

McCarty and McCarty also consider, in some detail, the types and kinds of postsynthetic modifications that can occur on histones, such as acetylation, methylation, and phosphorylation, which are reversible and which are often shown to be cell-cycle-related. Such an analysis points to the importance which these specific chemical modifications will have on the control of the cell cycle.

Gurley *et al.* (Ch. 3) consider extensively one of these postsynthetic modifications, namely, the phosphorylation of histones, especially histones H1 (interphase), $H1_M$ (mitosis), H2A, and H3, during the cell cycle. These authors conclude that phosphorylation of histone H3 and $H1_M$ could possibly be the driving force for chromatin condensation in the chromosome cycle, but that histone H1 (interphase) does not possess this function. This interphase type of phosphorylation of H1 has been proposed to be the mitotic "trigger" by Bradbury *et al.* (1,2). However, Gurley *et al.* feel that the interphase H1 is "involved in the formation or stabilization of the nucleofiliament." Phosphorylation of H2A, which occurs throughout the cell cycle, was envisioned as being necessary for maintenance of the heterochromatin.

Krause (Ch. 4), on the other hand, makes the point in her chapter that different histones can be selectively lost or retained during chromatin isolation at different phases of the cell cycle. This author finds that a selective leakage of histones from nuclei during isolation may account for discrepancies in earlier studies on the role of histone acetylation and phosphorylation in gene expression. Rather than ignore the selective leakage of certain histones, this phenomenon can be used advantageously as a way of studying the extent of histone modification in the various stages of genetic activation in chromatin. Among her conclusions are that H1 phosphorylation causes a loosening of nucleosomes in the 20-nm or native chromatin fiber and may, therefore, be correlated with increased transcriptional activity. Histone H3 is also reported to be more susceptible to dislodgement from transcriptionally active chromatin from cycling cells than from chromatin of noncycling cells.

Thus, three of these chapters suggest that histones and their postsynthetic modifications play an important role in cell replication, chromosomal condensation, and transcription. These authors make clear that the delineation of the exact roles of histones and of acidic (nonhistone) chromosomal proteins is far from complete. Nevertheless, the thrust of these chapters is to show that considerable progress has been made to further our understanding of the role of histones in chromatin structure and function.

It is readily appreciated by all investigators that every synthetic, enzymatic, and physiological process is sensitive to changes in hydrogen ion concentration. Although most would agree on the importance of intracellular pH as a potential regulator of intracellular processes, there has been a paucity of data on the subject due primarily to the difficulties encountered in the measurement of pH within a cell. To set the proper perspective, Gerson (Ch. 6) reviews the methods used to measure intracellular pH before proceeding to describe new findings on the role of intracellular pH changes as a function of the cell cycle of *Physarum*. The importance of intracellular pH as a regulator of mitosis is summarized, and it is noted that the rise in pH at the time of mitosis is, for example, correlated with the initiation of DNA synthesis in both mammalian cells and in *Physarum*. The chapter ends with a provocative, albeit speculative discussion on the significance of anomalies in hydrogen ion concentration as linked to cancer, which may be viewed as a cell-cycle-related disease. Gerson relates the onset of cancer to altered intracellular pH by an interesting review of the classical works of Warburg on aerobic glycolysis and lactic acid production in tumors, of Racker's ideas on altered intracellular pH in cancer, and of Burton's analysis of the incidence of cancer and altitute. A unifying and testable hypothesis on the role of intracellular pH and the cell cycle, thus, is provided. Hopefully, it will be applied to both cancer therapy and detection, and it will become a promising new approach to cell cycle regulation studies.

Harris (Ch. 5) has assembled a wealth of background information into a model on regulation of mitosis. Her chapter weaves together the diverse threads of information on the assembly of microtubules with an astute structural analysis on the regulatory role of calcium into a pattern for regulation of mitosis. Basically, her model emphasizes, rather than ignores, the role that the mass of vesicles of smooth endoplasmic reticulum in areas of microtubule assembly may have as an important element of control in the mitotic process. The ability of the membranous vesicles to release and sequester Ca^{2+} is presented in analogy to other motile systems, such as muscle, stalked ciliates, amoebae, as well as in acellular forms such as *Physarum.* In this model, onset of mitosis is dependent on the regulation of calcium concentration, which, in turn, affects several cellular processes: (a) microtubule assembly and disassembly, (b) contractive elements involving troponin, actin, and myosin, (c) subunit binding of phosphorylase kinase or other protein-binding sites (8), and (d) chromosome condensation (4). Heilbrunn's original emphasis on calcium as a regulator of cellular events is embodied in Harris's model, which may provide an insight into questions not explained by other current models of mitosis.

Cell cycle regulation in *Physarum polycephalum* is the subject of two chapters in this monograph (Gerson and Sauer). What makes *Physarum* such an interesting subject to study? It has a life cycle that makes it almost ideal for studies on regulation of the cell cycle and of cell differentiation. For example, during the plasmodial growth phase, up to 10^8 nuclei undergo synchronous mitosis at 8- to 10-hour intervals at room temperature. Techniques and procedures for growing and handling this organism during the entire life cycle have been published in considerable detail as noted in their bibliographies. Although a great deal is already known about the cell cycle of *Physarum,* we are now at the threshold of a better understanding of gene expression and cell cycle regulation, as demonstrated by the work of Sauer and Gerson. Sauer (Ch. 8) discusses quantitative and qualitative changes, which take place during the various phases of the cell cycle, in the flow of information from the nucleus to the rest of the cells. As previously mentioned, Gerson relates cell cycle changes in pH to the regulation of mitosis in this organism.

The chapter by Israel *et al.* (Ch. 10) on the regulation of expression of an inducible structural gene during the cell cycle of the green alga *Chlorella* presents evidence for a model of positive and negative oscillatory control of inducible gene regulation. Being a truly eukaryotic cell, this model highlights the mechanisms of regulation that have evolved from the prokaryotic level of organization. This chapter should serve to further detail the mechanism by which inducible enzymes are regulated in other nonphotosynthetic cell types.

Cell fusion as a means of studying cell cycle regulation is presented in the chapter by Rao and Sunkara (Ch. 7), using mammalian cells in culture.

This chapter deals primarily with G_1 phase of the mammalian cell cycle and the initiation of S phase. The authors conclude that inducers of S phase increase during the S phase itself and then decrease during the G_2 phase. Of particular importance is the fact that the inducer(s) of S is dependent, not on its concentration within the cell, but on its absolute quantity.

The regulation of protein synthesis during the cell cycle in *Chlamydomonas reinhardi* is examined in some detail by Howell and Baumgartel (Ch. 9). The authors consider the mechanisms whereby the overall rates of protein synthesis are controlled, as compared to the synthesis of single polypeptide species. Mindful of the difficulties inherent in measuring protein synthesis as a function of the cell cycle by analyzing the rate of incorporation of precursors against an unknown background of pool expansion, a new technique is presented whereby rates of polypeptide initiation and elongation can be measured independently of any changes in precursor pool sizes. As discussed by the authors, a change in the specific activity of differing classes of polyribosomes (with respect to size) can be taken as an indication of the rate of nascent protein chain growth. They also discuss the central fact that the rate of chain elongation varied during the cell cycle. Such an analysis is extended to an examination of the rate of aggregation or disaggregation of polysomes as a function of the cell cycle. The authors find that, indeed, there is a shift from the disaggregated to the aggregated form in relationship to the onset of the light and dark cycles that are used to induce synchrony in this organism. In fact, by a series of careful determinations, it is possible to determine the average size of the mRNA transcripts with the assumption that the average size of the translated mRNA is constant. The overall rate of protein synthesis during the cell cycle was determined and compared to the onset of increased amino acid incorporation. The authors conclude their analysis by discussing the relationship between changes in protein synthesis in the cell cycle in an attempt to see whether or not such changes are linked with the cytological control points in the cell cycle.

These authors next tackled the problem of determining the relative rates of labeling of individual membrane polypeptides and soluble polypeptides, as resolved by acrylamide gel electrophoresis. By careful autoradiography of these bands, the authors were able to determine which polypeptide species were maximally labeled at a specific time in the cell cycle. The authors discuss whether or not the patterns of maximal labeling for the membrane polypeptides versus the soluble polypeptides are phased with the light and dark periods of the cell cycle. With the exception of one soluble polypeptide, it is not clear to what extent illumination shifts are responsible for the periodic polypeptide labeling pattern. The authors conclude their presentation with a discussion of the relationship between the rates of polypeptide labeling and the transcriptional regulatory mechanisms that account for the differing rates of polypeptide chain initiation and elongation. An analysis of the kind presented by Howell and Baumgartel emphasizes the importance of translational mechanism controlling protein synthesis in the cell cycle and the types of experimentation that will allow future investigators to detect the points at which protein synthesis is controlled in the cell cycle.

The expression of cytoplasmic genes as a function of the cell cycle is examined in yeast by Johnson *et al.* (Ch. 11) by analyzing the pedigrees of a large number of individual yeast cells in which the *petites* are spontaneously induced and to see whether or not the frequency of *petite* production has any bearing toward the original paternal genomic characteristic. The authors then discussed the onset of asymmetry in the pedigrees in terms of the volume distribution of the population, which itself appeared to have a resemblance of the *petite* frequency distribution of the pedigree. The question of whether or not the amount of mitochondrial DNA is tightly coupled to the amount of nuclear DNA or that its synthesis is itself regulated at every cell cycle is discussed. The authors suggest that these quantities of DNA are not tightly coupled but, in fact, that changes in mitochondrial DNA are correlated with cell volume, perhaps in response to a requirement for greater mitochondrial function.

The authors consider other factors that may affect the asymmetry of production of *petites*, namely, the role of conjugation and the accompanying changes in cell volume during mating, and the role of the suppressive matings whereby the fates of diploid cells, which have been bequeathed a sufficient quantity of the positive mitgenomes, will give rise to different proportions of *petites* in their clones. The authors present what they call a differential cell cycle model to explain how the mitgenome population is modulated to account for the asymmetric pedigree (increasing incidence of *petites*). The authors examine the various parameters that affect conjugation as well as the expression of suppressives. This chapter emphasizes the value of a genetic approach to affect changes in the expression of cytoplasmic inheritable factors.

The last chapter of this book is on the regulation of epidermal mitotic activity in lower vertebrates. The chapter serves as a reminder that the regulation of cell cycle events in intact organisms occurs at various levels of biologic organization (from cellular to organismal). Hoffman (Ch. 12) reports that light serves as a dominant synchronizer of mitotic activity in lower vertebrates, just as it does in plants and algae such as *Chlamydomonas* and *Chlorella.* Synchrony of mitotic activity in photosynthetic cells has as its basis the periodic reception of light energy by the photosynthetic pigments. How then does light exert its synchronizing action in lower vertebrates? Is the light received by retinal, pineal, or some other photoreceptive elements, perhaps within their neural pathways? As yet there is no certain answer, for parapinealectomy and blinding abolish mitotic rhythm in the frog, *Rana pipiens,* whereas only continuous darkness, not blinding, arrested the mitotic rhythm in *Xenopus* larvae. Hoffman reports that some hypophyseal principle is required for epidermal proliferation. Prolactin, for instance, is shown to act as a stimulus to epidermal mitosis in fish, amphibians, and reptiles, but the stimulus may be indirect by stimulation of general metabolism, rather than specific to epidermis. Other hormones, such as thyroxine and corticoids, have also been implicated in the control of epidermal proliferation in the lower vertebrates. Hoffman concludes that squamates, amphibians, and reptiles are all affected by environmental factors such as light and that the effect is mediated via neurohypophyseal pathways.

The review by Hoffman demonstrates that the cell cycle in intact animals can be regulated by complex environmental mechanisms. In different studies, it has been shown that such extracellular mitotic controls can even act in different directions at the same time in an individual (3). For example, an individual with a rather large transplantable tumor may be in a nutritionally deficient state, and the resultant cancer-related weight loss induces in a number of cell populations, such as epidermis and gut epithelium, decreased cell proliferation. At the same time, however, the reticuloendothelial and lymphatic cell populations are shown to be undergoing increased cell proliferation, presumably due to the antigenic nature of the tumor and/or due to endotoxins derived from the tumor. Thus, the depleted nutritional status of the individual is apparently acting to decrease cell proliferation, in general, but the specific mitotic stimulus to the reticuloendothelial system cells is strong enough to prevail. It has been postulated that such a stimulus is brought about by specific receptors at the cell surface (6). It is not difficult to imagine that an epidermal or corneal wound could initiate a local increase of mitotic activity in cells immediately adjacent to the wound site as part of the healing response. This example illustrates that general, specific, and local regulation of cell cycle events may occur spontaneously in intact metazoans.

Clearly then, regulation of cell cycle events is a complex interaction between nuclear genome, cytoplasmic pools, and organelles, as well as the cell surface and its extracellular environment.

REFERENCES

1. Bradbury, E. M., Inglis, R. J., and Matthews, H. R., *Nature* (*London*) **247,** 257 (1974).
2. Bradbury, E. M., Inglis, R. J., Matthews, H. R., and Langan, T. A., *Nature* (*London*) **249,** 553 (1974).

3. Cameron, I. L., *in* "The Cell Cycle in Malignancy and Immunity" (J. C. Hampton, ed.), pp. 76–103. USERDA Technical Information Center, Oak Ridge, Tennessee, 1975.
4. Cameron, I. L., Sparks, R. L., Horn, K. L., and Smith, N. R., *J. Cell Biol.* (1977). **73,** 193.
5. Finch, J. T., and Klug, A., *Proc. Natl. Acad. Sci. U.S.A.* **73,** 1897 (1976).
6. Marx, J. L., *Science* **192,** 455 (1976).
7. Ris, H., *Ciba Found. Symp.* **28** [N.S.] (1975).
8. Weber, A., *Symp. Soc. Exp. Biol.* **30,** 445 (1976).

2

Some Aspects of Chromatin Structure and Cell-Cycle-Related Postsynthetic Modifications

Kenneth S. McCarty and Kenneth S. McCarty, Jr.

I. INTRODUCTION

We define the difference between prokaryotes and eukaryotes by the presence of chromatin, the macromolecular structure of which represents a highly organized complex of DNA and proteins confined to a nucleus. Histones constitute the major class of proteins that provide this delineation. Asymmetry, capacity to form spontaneous heterogenous interactions, and basicity of the amino-terminal residues provide the histones with a unique capacity to form a compact structural lattice on which the DNA is wound. All eukaryotes including yeasts, *Neurospora,* and *Physarum* (55, 69, 88, 103a, 145) demonstrate the presence of these basic proteins in concentrations that approximate the weight ratio of DNA per cell.

No other proteins studied, to date, have been shown to have a lower calculated mutation rate, which represents less than 0.06 residue changes per 100 total amino acid residues, over a period of a hundred million years (47). A striking example is the observation that only a 2% divergence in amino acid residues has been detected in histone class H4, even in species as widely separated as the pea and the calf (48).

Asymmetry and metabolic stability are two properties of histones that qualify them for providing the lattice on which DNA is wound with a compaction ratio of 6.8/1 (110). These properties emphasize the importance of the amino acid sequence essential for interaction with the surface of DNA, which results in an invariant structure that, once established, remains resistant to change.

A brief overall review of nucleosome structure and postsynthetic modifications is warranted in order to provide a framework on which to examine some of the subtle transitions temporally related to the various phases of the cell cycle. The magnitude of these facultative complex macromolecular structural alterations involved in condensed (heterochromatin) and extended (euchromatin) configurations is impressive, particularly as chromatin assumes its highly condensed form at metaphase.

II. CHROMATIN STRUCTURE

Recent studies using physical techniques include scanning electron microscopy (104), transmission electron microscopy (16, 108–112, 149), low angle neutron diffraction (14), circular dichroism (7, 13, 117, 162, 163), and other optical measurements. Ansevin (3–5) and Tsai (146) suggest a condensed form of DNA in chromatin. An examination of nucleolytic and/or proteolytic cleavage fragments agree with observations using nuclear magnetic resonance (10) to establish that the hydrophilic–cationic environment of histone amino terminal residues are involved in a structural association between these histone multimers and DNA in the formation of strong electrostatic interactions with the polyphosphate DNA backbone (Table I).

A. Physical Studies of the Nucleosome

Amino acid residues, such as proline, serine, and glycine, which favor extended chain structures, and basic amino acid residues, including lysine, arginine, and histidine, characterize the amino-terminal segments of histones, H2A, H2B, H3, and H4 (48). The central regions of H2A and H2B, as well as the carboxyl halves of H3 and H4 possess high proportions of apolar and other amino acid residues that appear to favor helix formation (45, 138). Thus, in general, aqueous solutions of histones are considered predominately to favor random coils, which are induced to interact with an increase in ionic strength. Under these conditions, well-defined structures predominate in the apolar regions to represent sites of histone–histone interactions. These hydrophobic interactions provide a configuration in which the amino-terminal residues have access to the aqueous environment, presumably providing primary sites of interaction of individual histones within the major groove of the DNA (103). When histones are dissolved in water, the CD spectrum resembles that of a random coil, and the fluorescence polarization of the tyrosines are low, indicating that these aromatic residues rotate freely (148). On the addition of salt, however, there is an immediate change in the CD and fluorescence anisotropy of all the histones, suggesting fewer degrees of freedom (27). Histones H3 and H4 demonstrate an additional slower change that takes place in minutes or hours, depending on the histone concentration (46). The fast step in CD change is best attributed to the formation of an α-helix, whereas the slow step is more likely to represent the formation of β-pleated sheets. Van Holde and Isenberg (147) propose that only limited histone sequences are involved in these molecular conformations. Histones in solution show specific cooperative folding that alter the degree of aggregation (147). Sedimentation equilibrium data show that $(H4–H3)_2$ represent a tetramer; H4–H2B and H2A–H2B represent dimers (45) in which H3–H4 and H2A–H2B have the highest binding constants. This undoubtedly explains the ease with which Kornberg and Thomas (78) were able to cross-link H2A–H2B. Some of these observations are outlined in Table I.

Physical properties of chromatin preparations that have been depleted of the lysine-rich histone H1 are all in agreement with the concept that the fundamental macromolecular structure is best represented by a chain of spherical particles ranging in size from 70 to 125 Å in diameter, connected by a flexible DNA filament (125). The spherical particles have been variously described as "nu" bodies (109, 110, 125), "nucleosomes" (114), or "ps particles," representing a single, double, or triple coil of native, perhaps "kinked" DNA (44, 135), wrapped around a hydrophobic histone domain composed of two dimers to include four classes of histones: H2A, H2B, H3, and H4 (45, 46). Thus, histone H1 is exceptional in that it does not represent a formal part of the histone–nucleosome complex. The basic unit of the chromatin structure has been visualized, using preparations of chromatin fibers spilled out of lysed nuclei (114). These structures resemble

TABLE I

Physical Techniques

Technique	Observation	Interpretation	Reference
Circular dichroism	↑ Ellipticity 260–300 m*M*	Urea affects protein secondary structure	(6, 7, 13, 39, 62, 63, 81, 82)
		Low ionic strength ↑ the effective (−) charge of the nucleoprotein backbone	
		Intermediate ionic strength weakens histone–DNA backbone	
Neutron diffraction	Concentration-dependent peak 10–11 nm	Interparticle space of a subunit	(11)
	Contrast match 5.5 and 3.7 nm difference from 11 nm	Histone and DNA have different spatial arrangement in subunit globular model in which apolar amino acid residues form a core surrounded by DNA complexed with basic amino-terminal residues	
	Radius of gyration of 185 bp give diameter 120–150 Å		
		Not all DNA is in an external shell	(14)
^{13}C Nuclear magnetic resonance spectra	Many narrow resonances are evident in histone spectrum	Propensity to form secondary structure at N terminal is low; therefore, free tails H4-H3 H2A-H4, H3, H2A, H3 attached to each other via hydrophobic interactions and each a tail for attachment to DNA	(86)
Proton magnetic resonance	C terminal H4 monomer PMR "invisible"	C terminal is relatively rigid secondary structure	

Nuclear magnetic resonance spectra	CNBr peptide Met84 to amino terminal versus carboxyl terminal	Flexible amino terminal and rigid carboxyl terminal	(10, 20, 21)
X-ray diffraction	Peak at 11 nm is concentration dependent	Neighboring subunits have a tendency to align themselves close to the axis of the fiber, or each subunit has three turns per 200 bp	(142, 148)
Electron microscopy	100 or 200–250 Å diameter fibers dependent on solvent condition	There may be considerable configurational heterogeneity in chromatin dependent on salt condition	(24, 25, 35, 36, 114)
		Freeze-fracture shows DNA enters and leaves at same site	
First derivative thermal denaturation	$T_{m_{III}} = 72°$ $T_{m_I} = 47°$	79 ± 3% DNA bp bound by histones	(3, 4, 5, 82, 146)
	$T_{m_{IV}} = 82°$ $T_{m_{II}} = 57°$	3.4 ± 0.4 amino acids per nucleotide	
Electron microscopy	Nucleosome structures 125 Å in diameter	Nucleosome spacing and size is a constant feature of chromatin preparations	(110, 114)
Neutron diffraction	10-nm peak has maxima which form a cross pattern with semimeridional angle 8°–9°	Coil of nucleosomes of pitch of 10 nm and outer diameter 30 nm	(7, 12–14, 24, 25, 37, 116, 117, 162)
	Low angle reflections 400, 200, and 140	Higher order structure pitch 500 Å radius 130 Å	
Centrifugation	Distribution of histones in alkali-denatured metrizamide gradients	Alkaline denaturation favors a bilateral model	(51, 52, 121)

those of preparations reconstituted from mixtures of DNA and the four histones added under specific solvation conditions and in proper sequence (19, 115, 159a).

B. Enzyme Probe Analysis of Nucleosome Structure

Enzymatic digestion of chromatin has been exceedingly useful in the analysis of the nucleosome. Proteases have been used to define histone–DNA interactions and nucleases to define the unit length of DNA. Some of these data and their interpretations are outlined in Table II.

The fact that trypsin digestion of intact nucleosomes is limited to 20–30 amino-terminal residues of H3, H4, H2B, and H2A (154) is in agreement with nuclear magnetic resonance studies (86) suggesting these basic amino acid residues as prime candidates to interact with the DNA. In addition, a more detailed analysis of the order of cleavage has been able to clarify a number of specific histone–DNA and histone–histone interactions (151). The observation that trypsin rapidly destroys histone H1 without disrupting the nucleosome subunit structure (150) is consistent with the concept that this histone is external to the nucleosome.

A number of nucleases have been used, both as a probe to analyze the organization of chromatin at the level of the basic repeating unit and as a technique to resolve the unit size of the DNA associated with the nucleosome.

Four classes of nucleases have been used to provide this information:

1. DNase I and DNase II have been shown to cleave a wide range of DNA sites on the chromatin substructure restricted by histone–DNA interactions.
2. The Ca^{2+}–Mg^{2+}-dependent endonucleases cleave those DNA sites that are regularly spaced along the repeating chromatin substructure without the release of mono- or oligonucleotides.
3. Micrococcal nuclease attacks the same sites as the Ca^{2+}–Mg^{2+} endonuclease, differing, however, in the release of mono- and oligonucleotides as acid soluble fragments.
4. Restriction enzymes as, for example, EcoRII, have been most useful to demonstrate that most, if not all, satellite and repetitive DNA's are confined to the nucleosome.

Early evidence of a subunit structure of chromatin using nucleases was first presented by Hewish and Burgoyne (61) as an extension of the work of Williamson (159). These studies, using DNase, show a 200 base pair limit digest fragment of DNA as a reflection of the regularity in the nucleotide chain length protected by the nucleosome protein–DNA interaction. Noll

TABLE II

Enzyme Probe Analysis of Nucleosome Structure

Enzyme	Observation	Interpretation	Reference
Trypsin	Digestion of only 20% of histone change in properties of nucleosome	Histones essential to maintain the compact structure	(40, 67, 68, 122, 123, 132, 143, 151)
	Cleavage of chromatin with trypsin cleaves only 20–30 amino acids H3, H4, H2B, H2A	Basic amino terminus defines coordinates of DNA binding sites and folding of DNA fiber in chromosome	
	Order of cleavage H1, H5, H3, H4, H2A, and H2B		
DNase I and II	DNA products from digestion form a regular series of fragment size classes, all being integral multimers of the smallest DNA unit	200 base pairs is a reflection of the regularity in distribution of protein	(2, 15, 28, 29, 105, 113, 124, 136, 152, 160, 161)
Staph nuclease	Digestion product of DNA of 185 and 140 base pairs	Limit digest of 140 bp	(8, 9, 42, 43, 137)
Ca^{2+}-Mg^{2+} endonuclease	Series of double-stranded fragments with single-stranded breaks	Double-stranded fragments result from a regular series of single strand fragments	(28, 60, 61, 131)
Micrococcal nuclease	Shows 3.4, 5.3, 8.6, 17, 22, and 26S fractions	Suggests either several types of subunits or asymmetric cleavages at any one of four or more sites	(50, 65, 116, 117, 127, 132, 133)
Restriction nuclease	Specific cleavage by EcoRIII	Repetitive DNA part of the nucleosome	(66, 119)

(105–107) obtained similar results when nuclei were treated with staphylococcal nuclease.

It is assumed that the nucleosome monomer excised by nuclease digestion is derived from the larger chromatin structure by clipping the DNA at relatively susceptible nucleotides between subunits. The detailed analysis of Sollner-Webb and Felsenfeld (137), using staphylococcal nuclease, however, shows a monomer DNA of 185 base pairs that is slightly heterogenous in size, presumably with tails of 45 base pairs. It is now generally accepted that the 140 base pair unit represents a limit digest as the most reasonable value for the DNA fragment wrapped around the nucleosome (Table II). It is presumed that these enzymatic probes reveal, in some way, the arrangement of histones with DNA wound on the exterior surface of the nucleosome subunit.

Finally, the combined use of trypsin and nuclease together, as seen in the work of Weintraub and Van Lente (154), has been valuable to show specific histone–DNA interactions. These studies of trypsin- and DNase-resistant DNA–histone fragments imply that both specific histone amino-terminal and carboxyl-terminal residues are associated with the DNase resistance. using cyanogen bromide cleavage of histone H4, it has been shown by Adler *et al.* (1) that the amino-terminal fragment 1–83 binds DNA with equal affinity as the whole molecule, whereas the C-terminal fragment (85–102) was weakly bound under the conditions tested.

C. Cross-Linking Probes of Nucleosome Structure

A more detailed analysis of the nucleosome has been undertaken using cross-linking experiments, as shown in Table III. These studies have already contributed a great deal to our knowledge of strong histone interactions with the histone octamer subunit.

Two cross-linking reagents, UV at 280 nm, and tetranitromethane may be considered as zero-length cross-linking reagents converting noncovalent interactions among the histones into covalent bonds, without the interposition of bridges (94). Both cross-linkers penetrate hydrophobic clusters and activate tyrosine residues (95). These studies show that histone H2B is cross-linked to both H2A and H4. Characterization of the cyanogen bromide peptide fragments shows that the amino-terminal half at tyrosine (Y) residue 40 of H2B is cross-linked to the amino-terminal half of histone H2A at residue phenylalanine (F) 25, whereas the carboxyl-terminal Y residue 121 of H2B is cross-linked to the F residue 61 of histone H4. In addition, it is most significant that two H2A molecules have been observed to be cross-linked to H2B, one of which has a methionine (Fig. 1A).

TABLE III

Cross-Linking Experiments

Techniques	Observation	Interpretation	Reference
UV	Covalent linkage of histones at 280 nm H2A–H2B	Histones must first interact with each other before being deposited on DNA	(95, 138–140)
	CNBr analysis of cross-linked histones	Histone H2B has separate binding sites for H2A and H4	(94)
Tetranitromethane	Cross-linked H2A–H2B	H4 is essential to proper reconstitution of H2A and H2B with DNA	(93)
		H4, H2A, and H2B interact specifically in chromatin	
Dimethyl suberimidate	$(H4)_2(H3)_2$ (H2A–H2B)	Tetramer of $(H4)_2(H3)_2$ aligomer of H2A-H2B monomer H1	(78)
Methyl-4-mercapto-butyrimidate	H3–H2B, H3–H4, H2B–H2A at high frequency	Proposes a model	(59)
Reporter dye	Does not interfere with histone binding	Histones in major groove	(54a)
Schiff base formation	Cross-link histone to DNA in nucleosome, 5′ DNA termini identified	Determined histone arrangement within nucleosome	(80, 134)
Formaldehyde	Formation of a reversible covalent bond between DNA and histone	Determine distribution of DNA on histones	(38, 59, 67, 68, 79)
Surface tension	F2B in the presence or absence of chromatin at ionic strength 0.01 and 0.10	Calculate amount of histone bound to DNA	(41)

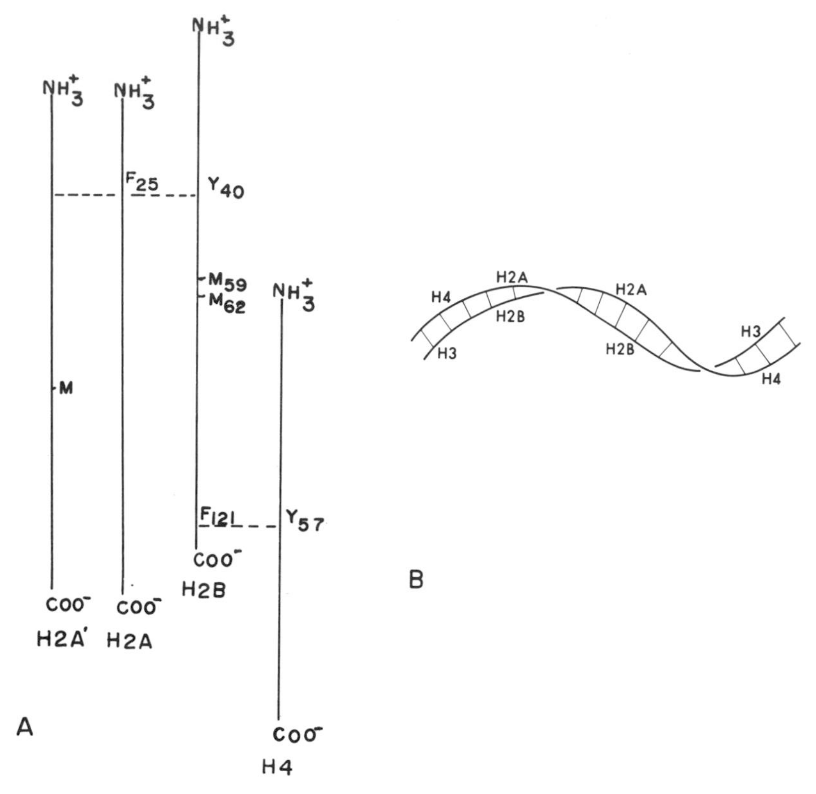

FIG. 1. (A) Results of UV cross-linking of nucleosomal histones. The specific residues were determined from an analysis of the cyanogen peptides (see text). Note that two forms of H2A–H2A′ are present within a single nucleosome (94, 95). (B) Positional association of histones with respect to the 140 base pair nucleosomal DNA determined by histone–DNA cross-linking, DNA fragment labeling (134).

Formaldehyde represents the shortest of the "spanner" cross-linking agents between lysine residues in chromatin. Using this agent, Van Lente (149a) has confirmed the observation that H2B is close to both H2A and H4. Both dimethylsuberimidate, 1-ethyl-3-(3-dimethylaminopropyl) carbodimide (78) and the sulfhydryl methylmercaptobuterimidate (59) shows, in addition, H3–H2B, H3–H4, and H2B–H2A are in close proximity.

The elegant work of Simpson (130, 131, 134) has recently revealed a number of fine details of the nucleosome structure, showing that the histones H3 and H4 interact at the ends of the nucleosomal DNA. In these studies, the DNA of the nucleosome core particle of 140 nucleotide base pair lengths is first labeled with [γ-^{32}P]ATP and polynucleotide kinase. The DNA is then methylated at a guanine residue, using dimethyl sulfate (103), gently depurinated, and cross-linked with histone in the formation of a Schiff base. The Schiff base is reduced with sodium borohydride to form a covalent linkage of histone and DNA. The DNA is then digested to permit the identification of a histone complex associated with the nucleotide as a marker. The most significant observations of these studies is the suggestion of symmetry in the arrangement of histones within the nucleosomes (Fig. 1B).

D. Immunochemical Probes of Nucleosome Structure

The nature of chromatin subunits have been examined, using two innovative techniques: immunosedimentation and immunoelectron microscopy (Table IV).

Antibodies to histone H1 have been useful to detect subtle irreversible changes as a result of the usual fractionation techniques, using acid or urea extraction (102). The use of anti-H2B–IgG molecules (134) has shown that the ratio of both H2A and (H2B + H3) to H4 are identical, suggesting that each nucleosome has an identical histone complement of two each of histones H2A, H2B, H3, and H4.

The information obtained by cross-linking experiments does not resolve the question of whether every nucleosome contains all four types of histones. Immunoelectron microscopy is a unique approach to this question. Bustin (31) has shown that 90% of the chromatin nucleosomes react with antibodies to histone H2B, suggesting that most of the chromatin nucleosomes contain this histone.

E. Histone H1

Only four of the major histone fractions are required to define the nucleosome core particle. These are the arginine-rich H3 and H4 and the moderately lysine-rich H2A and H2B. The H3 and H4 alone are sufficient to define the nucleosome when used in reconstruction experiments (155). Two molecules of each of the four histones are held together by hydrophobic interactions. As discussed above, the DNA fragment wound on the outside of the histone octamer is composed of 140 base pairs, and the DNA filament between nucleosomes represents a 40–50 base pair bridge.

The fifth histone, class H1, can be selectively extracted from the chromatin using salt concentrations in the range of 0.6 *M*. Upon removal of histone H1, the 40–60 base pair DNA bridge exhibits an increase in nuclease susceptibility (156). This is also accompanied by an increase in accessibility of the chromatin to DNA polymerase I (122). Both of these observations are in agreement with those of Bradbury (22) and others to

TABLE IV

Antihistone Antibodies

Techniques	Observation	Interpretation	Reference
Antibodies to histone and histone complexes	No cross-reaction between $(H3)_2$–$(H4)_2$ H2A–H2B, H1	Technique detected changes due to acid extraction	(30–34, 53, 56, 102, 118, 129, 141)
Immunosedimentation antibodies to histone	All nucleosomes contain H2B, and ratios of H2A, H4, and (H2B + H3) are same (±8%)	Each nucleosome identical histone complement; two each of histones H2A, H2B, H3, and H4	(134)
Immunoelectron microscopy of antibody–ferritin complex	All nucleosomes sediment at increased velocities	Useful technique for studying nucleosomes	(31, 134)

imply that histone H1 is complexed directly with DNA, providing a DNA–protein complex. This is in contrast to the hydrophobic protein–protein interactions proposed as the predominant force in the tetramers of $(H3–H4)_2$ and dimers of $(H2A)_2$ and $(H2B)_2$ within the nucleosome core.

The primary sequence of H1 has undergone extensive interspecies and interorgan variation as a result of both conservative and nonconservative amino acid replacements (48). The number of H1 subfractions varies from two to five. Although there is considerable variation in the amino acid residues 1–40 and 110–212, it is proposed that these sequences in the globular region are involved in the recognition of superhelical DNA requiring residues 72–106 (19, 137). Since the carboxyl terminal halves (72–212) of the histone H1 are enriched in lysine and proline, it is reasonable to presume that this represents the DNA-binding site. In contrast, the amino-terminal halves of these proteins are enriched in acidic residues and, therefore, might be expected to associate with components other than DNA. For example, two low molecular weight (MW) ($< 30,000$) proteins have been fractionated and characterized by their solubility in trichloroacetic acid. These have been termed "high-mobility group" (HMG) proteins; HMG1 and HMG2 and have been obtained in relatively pure form. From their amino acid sequences, these proteins are closely related (57, 128), found in large quantities, and seen to associate with histone H1. H1-b and H1-2 appear to interact only with the HMG1. Subfractions of histone H1, H1-3, and H1-3B bind both HMG1 and HMG2 (134a). These protein–protein interactions may account for some of the heterogeneity characteristic of histone H1.

F. Higher Orders of Nucleosome Packing

A number of recent experiments have demonstrated the key role played by H1 in higher orders of nucleosome packing and its temporal association with phosphorylation in the cell cycle (11, 23, 58, 64).

Finch and Klug (54) have shown that chromatin prepared by a brief digestion of nuclei with micrococcal nuclease and then extracted with 0.2 m*M* EDTA appear as loosely coiled filaments. In the presence of 0.2 m*M* Mg^{2+}, however, these nucleofilaments condense into a supercoil or "solenoidal structure" of a 10-nm pitch with six nucleosome particles per turn. It should be emphasized that the solenoidal structures require histone H1 for stabilization of higher orders of nucleosome packing (Fig. 2A).

Addition of histones H2A, H2B, H3, and H4 to SV40 DNA (3×10^6 daltons) under proper salt concentrations results in the formation of a 19–22 nucleosome "minichromosome" in the form of a DNA circle (150). If histone H1 is included, a contraction of the minichromosome results,

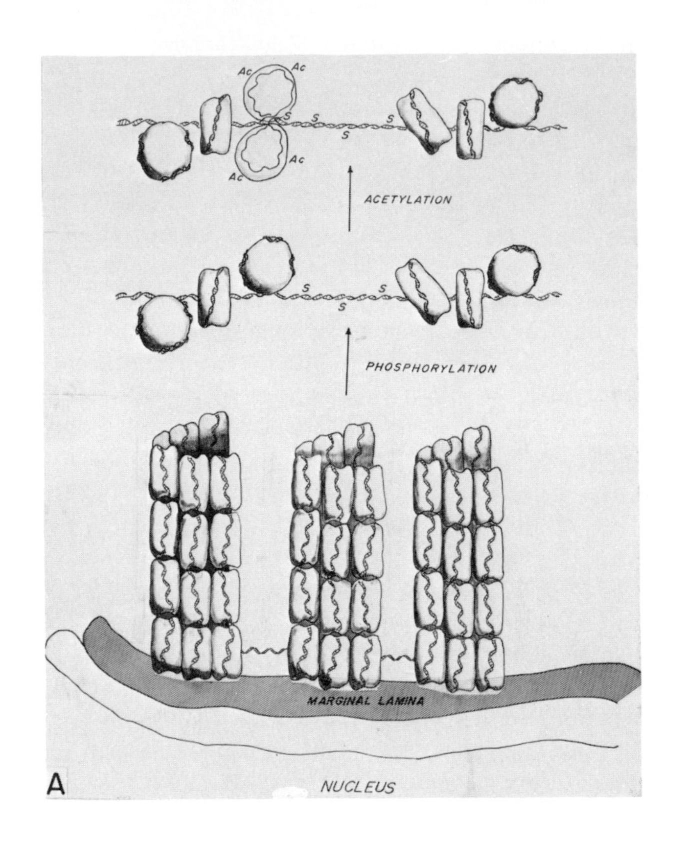
Ac
Ac
Ac
Ac
S
S
S
S
ACETYLATION
S
S
S
PHOSPHORYLATION
MARGINAL LAMINA
A
NUCLEUS

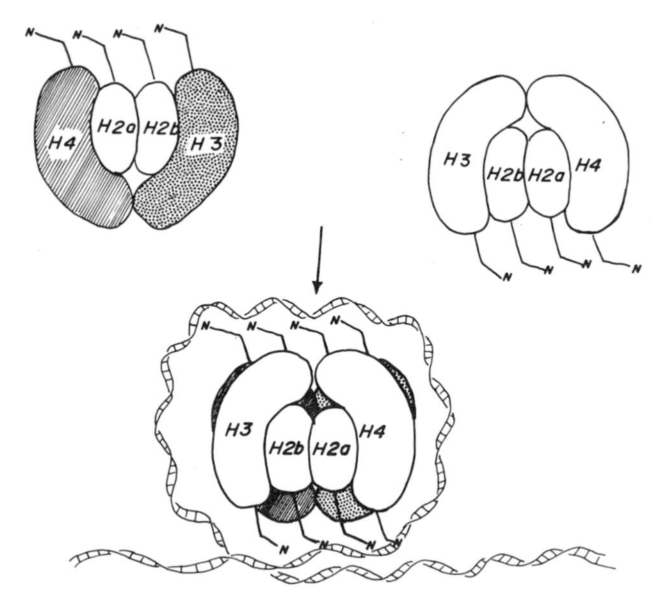
N
N
N
N
H4
H2a
H2b
H3
H3
H2b
H2a
H4
N
N
N
N
N
N
N
N
H3
H2b
H2a
H4
N
N
N
N
B

demonstrating morphological characteristics that resemble the solenoidal structure proposed by Finch and Klug (19a, 24–26, 54).

In the absence of histone H1 and/or magnesium, the repeat distance of the nucleosome and its DNA spacer is outlined in Table V. The coil length of the nucleosomal DNA is constant for all cells at 140 base pairs. In contrast, the spacer or bridge DNA varies with different cell types, usually shorter in active cell cultures (25 base pairs) and longer (60 base pairs) in resting liver cells. It is not possible to perceive the immediate significance of variations in spacer distance with active, or resting cell populations.

G. Nucleosome Subunit Conformation and RNA Transcription

Since the early proposal by Kossel (78a) and later by the Stedmans (140a) that histones may function to repress RNA transcription of DNA, a great deal of research has been directed to histone–DNA interactions. The main evidence for histone repression is based on numerous *in vitro* studies of RNA transcription using histone depleted chromatin (69a) and reconstructed chromatin, resulting in nucleosomal structures that resemble those seen in lysed nuclei (114). In these experiments, histones and DNA are first dissolved in 5 *M* urea and 2 *M* NaCl. It is essential to first remove the urea and then the 2 *M* NaCl to prevent nonspecific histone–DNA aggregation. Proof that the amino-terminal peptide sequences represent the potential site of repression of RNA transcription is given in experiments by Johns (69b). In these studies, only the amino-terminal cyanogen bromide fragment of histone H2B was capable of repression of RNA transcription using *Escherichia coli* polymerase.

We must pose the question then as to how the histone–DNA nucleosome complex functions in RNA transcription. The suggestion by Weintraub *et al.* (155) that the nucleosome may assume a transient open heterotypic dimer configuration is intriguing in that it has the potential to partially account for the function of the nucleosome structure. On the basis of the fact that most nucleosomes contain two each of the four histones H2A, H2B, H3, and H4, it may be presumed that there is a single pair of each of

FIG. 2. (A) Proposed conformations of nucleosomal degrees of packing associated with the cell cycle and transcription. The inactive histone H1 dephosphorylated highly compacted heterochromatic solenoidal nucleosomal structure is shown at the bottom figure associated with the marginal lamina. Phosphorylation of histone H1 partially relieves the high degree of compaction as shown in the middle diagram. This represents euchromatic histone H1 free non-permissive state. Acetylation is proposed as a requirement for active RNA transcription of a portion of the euchromatin as shown with one nucleosome in its half-nucleosome state. The process is reversible by deacetylation. (B) Molecular symmetry of the two heterotypic nucleosomal dimers and proposed DNA–amino-terminal histone residue interactions (155).

TABLE V

Relationship of DNA Spacer and Cell Type[a]

Cell	Total nucleosome repeat distance[b] (Base pairs)	Nucleosome core length[c] (Base pairs)	Internucleosome spacer distance[d] (K) (Base pairs)
Chick embryo	198	139	59
HeLa cells	183	135	48
Neuronal cells	170–175	140	30–35
Glial cells	195–200	140	55–60
Liver cells	200	140	60
CHO cells	165–178	140	25–38

[a] Lohr *et al.* (87a).
[b] Repeat distance = core length and spacer distance.
[c] DNA base pairs wound on the nucleosome.
[d] K = constant spacer distance for each cell type. Average multiple repeat distance = $140(n) + K(n - 1)$; n = number of nucleosomes.

the four histones per nucleosome. Weintraub and Groudine (152) suggest that all nucleosomes are composed of two isologous heterotypic protein tetramers (H2A, H2B, H3, and $H4)_2$. The DNA is presumed to cover a length of 95 base pairs in circumference, interacting with 10 base pairs per histone and a total length of 140 base pairs. Weintraub proposes that this structure opens up into separate "half-nucleosomes" permitting genetic readout. This mechanism is attractive in that it obviates the necessity of histone displacement. In this model, it is assumed that activation of the nucleosome represents its separation into a half-nucleosome configuration (Fig. 2B).

We would like to propose that the process of nucleosome unfolding is induced by a driving force represented by postsynthetic modifications of the histones, some of which will be briefly reviewed below. As an alternative to a complete scission of the nucleosome proposed by Weintraub, we suggest that it may also be conceivable that the nucleosome partially unfolds to provide the thermodynamic energy required for a partial separation of the DNA strands (92) (Fig. 2A and B).

III. TEMPORAL RELATION OF POSTSYNTHETIC MODIFICATIONS OF HISTONE AND GENE ACTIVITY

The real paradox concerning histones is that in spite of numerous molecular constraints (including sequence conservation, limited diversity in the number of classes, constant protein/DNA ratio, and metabolic stability), as many as 240 different molecules have been resolved, all of which demonstrate the same amino acid composition as the parent unmodified histone (48). These molecules represent the consequence of both single and multiple postsynthetic modifications to include: acetylation (49, 96), methylation (18), phosphorylation (58, 58a, 75, 76, 97, 101), poly-ADP-ribosylation (71–74, 144), etc. It should be emphasized that all of these are reversible (120), some hormone dependent (77, 84, 85, 87, 89–91, 98), and many cell cycle related (17, 23).

As shown in Table VI, gene activation of each of the histones may be correlated with numerous postsynthetic modifications. Two cardinal features are (1) the fact that a limited number of residues are involved, and (2) that this involvement is limited to those amino terminal residues most likely to represent charge interactions with the DNA nucleosome (Table VII).

As discussed by Gurley (58) in this volume, phosphorylation has been temporally correlated with the cell cycle, particularly histones H1 and H3 (64). In the mammary gland culture, induction of cell division by insulin also results in the phosphorylation of H2A (91) in serine residue 19. Once again, this emphasizes that only a limited number of specific amino acid residues are involved (Table VIII).

Phosphorylation and acetylation represent postsynthetic modifications that take place in both the cytoplasm and nucleus (Table VII). For example, as histone H4 mRNA is translated, the amino-terminal serine is both phosphorylated and acetylated (120). An additional lysine residue is acetylated prior to its transport to the nucleus. A number of nuclear modifications include acetylation of lysine 5, 8, 12, and 16 (Table VII). Methylation of lysine 20 and phosphorylation of a histidine residue (41a) represent other nuclear events.

It should be emphasized that all of these modifications of histone H4 are rapidly reversible, with the exception of acetylation of the amino-terminal

TABLE VI

Temporal Relation of Postsynthetic Modifications and Gene Activity

Histone	Modification	Activity	References
H3, H4	Acetylation	↑ RNA synthesis lymphocytes	(70)
H3, H4	Deacylation	(−) RNA synthesis *Arbacia lixula*	(120)
H3	Acetylation	↑ RNA synthesis HeLa cells	(99, 158)
H3	Phosphorylation	Mitosis	(21)
H2A	Phosphorylation	↑ RNA mammary gland	
H1	Phosphorylation		(21)

TABLE VII

Residues Involved in Histone Postsynthetic Modifications[a]

Histone	Phosphorylation		Acetylation		Methylation		Reference
	Cytoplasm	Nucleus	Cytoplasm	Nucleus	Cytoplasm	Nucleus	
H1	—	$S_{37,105,160,180}$	—	—	—	—	(79a)
		$T_{16,136}R$, K_{212}					(144a, 144b)
H2A	S_1	S_{19}	S_1	K_5	—	—	(101)
H2B	—	$S_{6,32,36,38}$	—	$K_{5,12,15,20}$	—	—	(144a)
H3	—	$S_{10,28}$	—	$K_{9,14,18,23}$	—	K_9, K_{27}	(99, 144a)
H4	S_1	—	S_1	$K_{5,8,12,16}$	—	K_{10}, K_{20}	(88a, 88b, 144a)
H5	—	S	—	—	—	—	(144c)
H6	—	S_5	—	—	—	—	(144a)

[a] The letters refer to the amino acids that are modified, using the one-letter code designation for the amino acids. S, serine; T, threonine; R, arginine; K, lysine. The subscript designates the specific residue that it modifies when known. Specifically, in the case of histone H2A and H4, the modification takes place in the cytoplasm, as well as the nucleus on the amino terminal series.

TABLE VIII

Amino Acid Sequences Surrounding Sites of Phosphorylation by Protein Kinases[a]

Protein modified	Sequence	Origin of enzyme	Reference
Histone H1	R-(K)-A-[$\overset{37}{S}$]-G-P	Rat liver, calf liver, pig brain	(79a)
	S-G-[$\overset{105}{S}$]-F-(K)-L	Rat liver, calf liver, pig brain	
	$\genfrac{}{}{0pt}{}{R}{K}$-K-[$\overset{160}{S}$]-P-(K)	Trout testis	(144a)
Histone H2A	N-Ac-[$\overset{1}{S}$]-G-(R)-G	Trout testis	(144a)
	T-(R)-S-[$\overset{19}{S}$]-R-A	Mouse mammary gland	(89)
Histone H2B	R-(K)-E-[$\overset{36}{S}$]-T-S-V	Pig brain, lymphocyte	(68a)
	K-(K)-G-[$\overset{14}{S}$]-K-A	Pig brain, lymphocyte	(126)
	P-A-$\overset{\text{Ac 6}}{K}$-[S]-A-P-(K)	Trout testis	(49)
Histone H4	N-Ac-[$\overset{1}{S}$]-G-(R)-G-K	Trout testis	(49)
Protamines	(R)-R-[S]-S-S-R-P	Trout sperm	(49)
	(R)-V-[S]-R-(R)	Trout sperm	
Phosphorylase B	Q-I-[S]-$\genfrac{}{}{0pt}{}{V}{I}$-(R)-G	Rabbit, human muscle	(104a)

[a] Serine residue bracket indicates locus of phosphorylation. Superscript indicates residue. A = alanine; E = glutamic; G = glycine; I = isoleucine; K = lysine; L = leucine; P = proline; R = argenine; S = serine; V = valine. Other abbreviations as in Table VII. Numbers represent residue number modified

e.g., $\overset{38}{S}$ = serine residue 38

by phosphorylation. N = amino terminal; Ac = acetylated residue.

TABLE IX

Methylation of Histones[a]

	Methyltransferase		
Enzyme	I	II	III
Location	Cytosol	Cytosol	Nuclei
Residue methylated	Arginine	Glu and Asp	Lys
Protein	Histones	Gelatin	Histones
	Mono R	Ovalbumin	Mono K
	Dimer R	Ovalbumin	Di K
		RNase	Tri K
	H3–H4 > H2B–H2A		H3–H4 > H2B–H2A
	None H1		
	H3 Mono R and H		
Amount	(80–100%)		(80–100%)
Time of cell cycle	(Max) G_2 and M		(Max) G_2 and M
Demethylation			ε-Alkyllysinase

[a] Paik and Kim (114a).

serine. The acetylation of the amino-terminal serine of both histone H4 and H2A are cytoplasmic events that take place only during the S phase of the cell cycle and is essentially irreversible (100).

The most significant observation is that all postsynthetic modifications appear to take place on the amino-terminal histone residues (1–40) and, therefore, would presumably have the capacity to alter the charge interaction with DNA. The process of acetylation utilizes both acetyl-CoA and the transferase enzymes firmly bound to chromatin, as are both the protein kinases and methylases. Methylation utilizes *S*-adenosylmethionine.

Specific histone residues that are phosphorylated are presented in Table VIII, and those that are methylated are shown in Table IX.

IV. DISCUSSION

A consideration of the number of precisely ordered individual events essential to the control of cell replication leaves little doubt of its incredible complexity. In an attempt to limit the magnitude of the problem, this review has emphasized the evidence of some of the basic features of the nucleosome as representative of the major structural unit of chromatin. The evidence for this is the fact that nucleosomal DNA represents 50–75% of the total DNA of the genome. If the nucleosome is to be accepted as the ubiquitous component of both euchromatin and heterochromatin, its relation to RNA transcription and the cell cycle should be considered. Four immediate questions come to mind: (a) What are the driving forces involved in the formation of the solenoidal superstructure in mitosis? (b) Assuming that the 140 base pair nucleosomal DNA is inaccessible to endonucleases (40-Å gyration radii), how is it possible to transcribe DNA utilizing an RNA polymerase with similar or even greater dimensions? (c) Since the nucleosomal DNA represent only 140 base pairs, it is insufficient to code for even the smallest histone H4 (102 amino acids or 306 base pairs). Thus we must conclude that most genomes are involved with more than one nucleosome, thereby necessitating transcription through multiple nucleosomes. (d) What aspects of the chromatin structure function in determining the high degree of selectivity in limiting RNA transcription to 3–10% of the total genetic material of the eukaryote?

Although significant morphological changes occur within the chromatin fiber during mitosis, the basic nucleosome subunit structure appears to be retained by the mitotic chromosome (157). This observation also emphasizes the observed stability of histone–histone and histone–DNA interactions during mitosis. The explanation for the additional chromatin condensation must be sought in minor postsynthetic modifications that alter molecular associations. A prime candidate for this role is the histone H1 (58), particularly since it is required for higher order nucleosome packing ratios, or solenoidal structures.

The packing order of the heterochromatin in the metaphase chromosome is maximal, representing a solenoidal superstructure held together by histone H1. Postsynthetic modifications of this histone have been proposed to influence chromatin condensation. The molecular mechanism of DNA replication controlled by phosphorylation (58), however, is not immediately evident at this time. As seen in Table X, specific histone H1 residues are phosphorylated in G_1, S, and mitosis, with a total of one, three, and six modifications.

In conclusion, we would like to propose as a working hypothesis that postsynthetic modifications of histones alter the capacity of the DNA genome template to be transcribed and/or replicated. This model, Fig. 2A, suggests alterations of protein–DNA affinities that dictate the stage of nucleosome aggregation. For example, the condensed inactive solenoidal heterochromatin structure is dictated by the presence of histone H1 and modulated by phosphorylation during the cell cycle. We suggest further that euchromatin exists in two conformational states, inactive-permissive heterotypic dimer with two copies of each of four histones which can be induced to form a half nucleosome with one copy of each of four histones (153). We propose that acetylation plays an important role in this modification. Thus a steroid receptor could be responsible, as a consequence of protein-protein interactions (acetylase–receptor complex) in the activation of specific nucleosomes in the formation of active half nucleosomes (98). New

TABLE X

Phosphorylation and Cell Cycle

Histone	Cell cycle phase	Residues phosphorylated			References
		Ser	Thr	Total	
H1	Late G_1	105	—	1	(64, 79a)
	S	105, 160, 180	—	3	(64, 79a)
	Mitosis	37, 105, 160, 180	16, 136	6	(64, 79a)
H2A	Late G_1	19	—	1	(91, 120)
	S	1, 19	—	2	
H3	Mitosis	10	—	1	(99)
	Interphase	28	—	1	(99)

[a] Phosphorylation has been shown to be related to the phase of the cell cycle. For example, histone H1 is phosphorylated in specific residues in late G_1, S, and mitosis. H2A is phosphorylated in late G_1, the mammary gland is phosphorylated on the amino-terminal series at the time that it is synthesized. H3 is phosphorylated during mitosis, and there is evidence that the residue that is phosphorylated is serine (58a).

evidence for this model has recently been suggested by experiments which demonstrate selective DNase II sensitivity for active nucleosomes (152).

V. SUMMARY

It is now generally accepted that the nucleosome is an essential structural component on which the DNA is organized. Its role in replication and transcription, however, is far from clear. Intuitively, it is reasonable to suggest that a driving force for these events is likely to be mediated by postsynthetic modifications. Before this concept will be acceptable, however, a great deal remains to be done to define the process and to elucidate the exact role of both histones and the acidic chromosomal proteins.

ACKNOWLEDGMENTS

Supported by NIH Grants 5R01-GM-12805-10, G-72-3857 and CB-63996-34 from the National Cancer Institute.

REFERENCES

1. Adler, A. J., Fulmer, A. W., and Fasman, G. D., *Biochemistry* **14,** 1445 (1975).
2. Altenburger, W., Hörz, W., and Zachau, H. G., *Nature* (*London*) **264,** 517 (1976).
3. Ansevin, A. T., and Brown, B. W., *Biochemistry* **10,** 1133 (1971).
4. Ansevin, A. T., Hnilica, L. S., Spelsberg, T. C., and Kehm, S. L., *Biochemistry* **10,** 4793 (1971).
5. Ansevin, A. T., MacDonald, K. K., Smith, C. E., and Hnilica, L. S., *J. Biol. Chem.* **250,** 281 (1975).
6. Augenlicht, L. H., and Lipkin, M., *Biochem. Biophys. Res. Commun.* **70,** 540 (1976).
7. Augenlicht, L. H., Nicolini, C., and Baserga, R., *Biochem. Biophys. Res. Commun.* **59,** 920 (1974).
8. Axel, R., *Biochemistry* **14,** 2921 (1975).
9. Axel, R., Melchior, W., Jr., Sollner-Webb, B., and Felsenfeld, G., *Proc. Natl. Acad. Sci. U.S.A.* **71,** 4101 (1974).
10. Azizova, O. A., Bushuev, V. N., Korol, V. A., Kayushin, L. P., Sibel'dina, L. A., Nikolaev, Yu. V., and Kopilov, V. A., *Biofizika* **20,** 327 (1975).
11. Baldwin, J. P., Boseley, P. G., Bradbury, E. M., and Ibel, K., *Nature* (*London*) **253,** 245 (1975).
12. Bartley, J. A., and Chalkley, R., *J. Biol. Chem.* **247,** 3647 (1972).
13. Bartley, J. A., and Chalkley, R., *Biochemistry* **12,** 468 (1973).
14. Baudy, P., Bram, S., Vastel, D., and Lepault, J., *Biochem. Biophys. Res. Commun.* **72,** 176 (1976).
15. Billing, R. J., and Bonner, J., *Biochim. Biophys. Acta* **281,** 453 (1972).
16. Blyudenov, M. A., *Tr. Prikl. Bot., Genet. Sel.* **52,** 29 (1973).

17. Borun, T. W., Paik, W. K., Lee, H. W., Pearson, D., and Marks, D. B., *Cold Spring Harbor Conf. Cell Proliferation,* **1,** p. 701 (1974).
18. Borun, T. W., Paik, W. K., Lee, H. W., Pearson, D., and Marks, D.B., *in* "Control of Proliferation in Animal Cells" (B. Clarkson and R. Baserga, eds.), p. 701. Cold Spring Harbor Lab., Cold Spring Harbor, New York, 1974.
19. Boseley, P. G., Bradbury, E. M., Butler-Browne, R. S., Carpenter, B. G., and Stephens, R. M., *Eur. J. Biochem.* **62,** 21 (1976).
19a. Bottger, M., Scherneck, S., and Fenske, H., *Nucleic Acids Res.* **3,** 419 (1976).
20. Bradbury, E. M., Carpenter, B. G., and Rattle, H. W. E., *Nature* (*London*) **241,** 123 (1973).
21. Bradbury, E. M., Cary, P. D., Crane-Robinson, C., Rattle, H. W. E., Boublik, M., and Sautière, P., *Biochemistry* **14,** 1876 (1975).
22. Bradbury, E. M., Danby, S. E., Rattle, H. W. E., and Ginacotti, V., *Eur. J. Biochem.* **57,** 97 (1975).
23. Bradbury, E. M., Inglio, R. J., and Matthews, H. R. *Nature* (*London*) **247,** 257 (1974).
24. Bram, S., Butler-Browne, G., Baudy, P., and Ibel, K., *Proc. Natl. Acad. Sci. U.S.A.,* **72,** 1043 (1975).
25. Bram, S., *Biochimie* **57,** 1301 (1975).
26. Bram, S., Butler-Browne, G., Morton Bradbury, E., Baldwin, J. P., Reiss, C., and Ibel, I., *Biochimie* **56,** 987 (1974).
27. Brodie, S., Giron, J., and Latt, S. A., *Nature* (*London*) **253,** 470 (1975).
28. Burgoyne, L. A., Hewish, D. R., and Mobbs, J. D., *Biochem. J.* **143,** 67 (1974).
29. Burgoyne, L. A., Mobbs, J. D., and Marshall, A. J., *Nucleic Acids Res.* **3,** 3293 (1976).
30. Bustin, M., *Nature* (*London*), **245,** 207 (1973).
31. Bustin, M., Goldblatt, D., and Sperling, R., *Cell* **7,** 297 (1976).
32. Bustin, M., and Kupfer, H., *Biochem. Biophys. Res. Commun.* **68,** 718 (1976).
33. Bustin, M., Rall, S. C., Stellwagen, R. H., and Cole, R. D., *Science* **163,** 391 (1969).
34. Bustin, M., Yamasaki, H., Goldblatt, D., Shani, M., Huberman, E., and Sachs, L., *Exp. Cell Res.* **97,** 440 (1976).
35. Carlson, R. D., and Olins, D. E., *Nucleic Acids Res.* **3,** 89 (1976).
36. Carlson, R. D., Olins, A. L., and Olins, D. E., *Biochemistry* **14,** 3122 (1975).
37. Carpenter, B. G., Baldwin, J. P., Bradbury, E. M., and Ibel, K., *Nucleic Acids Res.* **3,** 1739 (1976).
38. Chalkley, R., and Hunter, C., *Proc. Natl. Acad. Sci. U.S.A.* **72,** 1304 (1975).
39. Chang, C., Weiskopf, M. H., and Li, H. J. *Biochemistry* **12,** 3028 (1973).
40. Chatterjee, S., and Walker, I. O., *Eur. J. Biochem.* **34,** 519 (1973).
41. Chattoraj, D. K., Bull, H. B., and Chalkley, R., *Arch. Biochem. Biophys.* **152,** 778 (1972).
41a. Chen, C.-C., Smith, D., Bruegger, B., Halpern, R., and Smith, R., *Biochemistry* **13,** 3785 (1974).
42. Clark, R. J., and Felsenfeld, G., *Nature* (*London*), *New Biol.* **229,** 101 (1971).
43. Clark, R. J., and Felsenfeld, G., *Biochemistry* **13,** 36222 (1974).
44. Crick, F. H. C., and Klug, A. *Nature* (*London*) **255,** 530 (1975).
45. D'Anna, J. A., Jr., and Isenberg, I., *Biochem. Biophys. Res. Commun.* **61,** 343 (1974).
46. D'Anna, J. A., Jr., and Isenberg, I., *Biochemistry* **13,** 4987 (1974).
47. Dayhoff, M. O., ed., "Atlas of Protein Sequence and Structure, Vol. 4, p. 42. Nat. Biomed. Res. Found., Washington, D.C., 1969.
48. DeLange, R. J., and Smith, E. L., *Ciba Found. Symp.* **28** (*new ser.*), 59 (1975).
49. Dixon, G. H., Candido, E. P. M., Honda, B. M., Louie, A. J., MacLeod, A. R., and Sung, M. T., *Ciba Found. Symp.* **28** (*new ser.*), 229 (1975).

50. Doenecke, D. *Cell* **8,** 59 (1976).
51. Doenecke, D., and McCarthy, B. J., *Biochemistry* **14,** 1366 (1975).
52. Doenecke, D., and mcCarthy, B. J., *Biochemistry* **14,** 1373 (1975).
53. Elgin, S. C. R., and Boyd, J. B., *Chromosoma* **51,** 135 (1975).
54. Finch, J. T., and Klug, A., *Proc. Natl. Acad. Sci. U.S.A.* **73,** 1897 (1976).
54a. Gabbay, R., DeStefano, R., and Sanford, K., *Biochem. Biophys. Res. Commun.* **46,** 155 (1972).
55. Goff, C. G., *J. Biol. Chem.* **251,** 4131 (1976).
56. Goldblatt, D., and Bustin, M., *Biochemistry* **14,** 1689 (1975).
57. Goodwin, G. H., Nicolas, R. H., and Johns, E. W., *FEBS Lett.* **64,** 412 (1976).
58. Gurley, L. R., Tobey, R. A., Walters, R. A., Hildebrand, C. E., Hohmann, P. G., D'Anna, J. A. Jr., Barham, S. S., and Deaven, L. L. Chapter 3, this volume (1977).
58a. Gurley, L. R., Walters, R. A., and Tobey, R. A. *J. Biol. Chem.* **250,** 3936 (1975).
59. Hardison, R. C., Eichner, M. E., and Chalkley, R., *Nucleic Acids Res.* **2,** 1751 (1975).
60. Hewish, D. R., *Nucleic Acids Res.* **3,** 69 (1976).
61. Hewish, D. R., and Burgoyne, L. A., *Biochem. Biophys. Res. Commun.* **52,** 504 (1973).
62. Hjelm, R. P., Jr., and Huang, R. C. C., *Biochemistry* **13,** 5275 (1974).
63. Hjelm, R. P., Jr., and Huang, R. C. C., *Biochemistry* **14,** 2766 (1975).
64. Hohmann, P. G., Tobey, R. A., and Gurley, L. R., *J. Biol. Chem.* **251,** 3685 (1976).
65. Honda, B. M., Baillie, D. L., and Candido, E. P. M., *J. Biol. Chem.* **250,** 6443 (1975).
66. Hörz, W., Igo-Kemenes, T., Pfeffer, W., and Zachau, H. G., *Nucleic Acids Res.* **3,** 3213 (1976).
67. Hyde, J. E., and Walker, I. O., *FEBS Lett.* **50,** 150 (1975).
68. Hyde, J. E., and Walker, I. O., *Nucleic Acids Res.* **2,** 405 (1975).
68a. Iwai, K., Hayashi, H., and Ishikawa, K., *J. Biochem.* (*Tokyo*) **72,** 357 (1972).
69. Jerzmanowski, A., Staron, K., Tyniec, B., Bernhardt-Smigielska, J., and Toczko, K., *FEBS Lett.* **62,** 251 (1976).
69a. Johns, E. W., and Forrester, S., *Biochem. J.* **111,** 371 (1969).
69b. Johns, E. W., and Hoare, T., *Nature* (*London*) **226,** 650 (1970).
70. Johnson, E. M., and Allfrey, V. G. *Arch. Biochem. Biophys.* **152,** 786 (1972).
71. Johnson, E. M., and Hadden, J. W., *Science* **187,** 1198 (1975).
72. Johnson, E. M., Hadden, J. W., Inoue, A., and Allfrey, V. G., *Biochemistry* **14,** 3873 (1975).
73. Johnson, E. M., Inoue, A., Crouse, L. J., Allfrey, V. G., and Hadden, J. W., *Biochem. Biophys. Res. Commun.* **65,** 714 (1975).
74. Johnson, E. M., Karm, J., and Allfrey, V. G., *J. Biol. Chem.* **249,** 4990 (1974).
75. Kang, Y.-J., Olson, M. O. J., Jones, C., and Busch, H., *Cancer Res.* **35,** 1470 (1975).
76. Karn, J., Johnson, E. M., Vidali, G., and Allfrey, V. G., *J. Biol. Chem.* **249,** 667 (1974).
77. Keller, R. K., Socher, S. H., Krall, J. F., Chandra, T., and O'Malley, B. W., *Biochem. Biophys. Res. Commun.* **66,** 453 (1975).
78. Kornberg, R. D., and Thomas, J. O., *Science* **184,** 865 (1974).
78a. Kossel, A., "The Protamines and Histones." Longmans, Green, New York, 1928.
79. Kozlov, Yu. I., and Debabov, V. G., *Mol. Biol.* **7,** 242 (1973).
79a. Langan, T. A., *Ann. N.Y. Acad. Sci.* **185,** 166 (1971).
80. Levina, E. S., and Mirzabekov, A. D., *Dokl. Akad. Nauk SSSR* **221,** 1222 (1975).
81. Li, H. J., *Nucleic Acids Res.* **2,** 1275 (1975).
82. Li, H. J., Chang, C., Evagelinou, Z., and Weiskopf, M. H., *Biopolymers* **14,** 211 (1975).
83. Li, H. J., and Flemly, D. A., *Anal. Biochem.* **52,** 300 (1973).
84. Libby, P. R., *Biochem. Biophys. Res. Commun.* **31,** 59 (1968).
85. Libby, P. R., *Biochem. J.* **134,** 907 (1973).

86. Lilley, D. M. J., *Proc. Conf. Mol. Spectrosc., 6th,* (1977).
87. Liu, A. Y., and Greengard, P., *Proc. Natl. Acad. Sci. U.S.A.* **73,** 568 (1976).
87a. Lohr, D., Corden, J., Tatchell, K., Kovacic, R. T., and Van Holde, K. E. *Proc. Natl. Acad. Sci. U.S.A.* **74,** 79 (1977).
88. Lohr, D., and Van Holde, K. E., *Science* **188,** 165 (1975).
88a. Louie, A.J., and Dixon, G.H. *Proc. Natl. Acad. Sci. U.S.A.* **69,** 1975 (1972).
88b. Louie, A.J., and Dixon, G.H. *Nature* (*London*) **243,** 164 (1973).
88c. Louie, A.J., Sung, M., and Dixon, G. H. *J. Biol. Chem.* **248,** 3335 (1973).
89. McCarty, K. S., and McCarty, K. S., Jr., *J. Natl. Cancer Inst.* **53,** 1509 (1974).
90. McCarty, K. S., and McCarty, K. S., Jr., *J. Dairy Sci.* **58,** 1022 (1975).
91. McCarty, K. S., and McCarty, K. S., Jr., *in* "Control Mechanisms in Cancer" (W. E. Criss, T. O. Ono, and J. R. Sabine, eds.), p. 37. Raven, New York, 1976.
92. McCarty, K. S., Jr. and McCarty, K. S. *Am. J. Pathol.* **86,** 703 (1977).
93. Martinson, H. G., and McCarthy, B. J., *Biochemistry* **14,** 1073 (1975).
94. Martinson, H. G., and McCarthy, B. J., *Biochemistry* **15,** 4126 (1976).
95. Martinson, H. G., Shetlar, M. D., and McCarthy, B. J., *Biochemistry* **15,** 2002 (1976).
96. Marushige, K., *Proc. Natl. Acad. Sci. U.S.A.* **73,** 3937 (1976).
97. Marushige, K., Ling, V., and Dixon, G. H., *J. Biol. Chem.* **244,** 5953 (1969).
98. Marzluff, W. F., and McCarty, K. S., *J. Biol. Chem.* **245,** 5635 (1970).
99. Marzluff, W. F., and McCarty, K. S., *Biochemistry* **11,** 2672 (1972).
100. Marzluff, W. F., Sanders, L. A., Miller, D. M., and McCarty, K. S., *J. Biol. Chem.* **247,** 2026 (1972).
101. Gallinaro-Matringe, H. G., and Jacob, J., *FEBS Lett.* **36,** 105 (1973).
102. Mihalakis, N., Miller, O. J., and Erlanger, B. F., *Science* **192,** 469 (1976).
103. Mirzabekov, A. D., and Melnikova, A. F., *Mol. Biol. Rep.* **1,** 379 (1974).
103a. Morris, N., *Cell* **8,** 357 (1976).
104. Nakanishi, Y. H., Kawabara, M., and Nei, T., *J. Electron Microsc.* **24,** 123 (1975).
104a. Neurath, H., and Hill, R. L., eds., "The Proteins," 3rd ed., Vols. 1 and 2. Academic Press, New York, 1975 and 1976 resp.
105. Noll, M., *Nature* (*London*) **251,** 249 (1974).
106. Noll, M., *Nucleic Acids Res.* **1,** 1573 (1974).
107. Noll, M., Thomas, J. O., and Kornberg, R. D., *Science* **187,** 1203 (1975).
108. Olins, A. L., Carlson, R. D., and Olins, D. E., *J. Cell Biol.* **64,** 528 (1975).
109. Olins, A. L., Carlson, R. D., Wright, E. B., and Olins, D. E., *Nucleic Acids Res.* **3,** 3271 (1976).
110. Olins, A. L., and Olins, D. E., *Science* **183,** 330 (1974).
111. Olins, D. E., and Olins, A. L., *J. Cell Biol.* **53,** 715 (1972).
112. Olins, D. E., and Olins, A. L., *Symp. Br. Soc. Cell Biol., 1975* (1975).
113. Oosterhof, D. K., Hozier, J. C., and Rill, R. L., *Proc. Natl. Acad. Sci. U.S.A.* **72,** 633 (1975).
114. Oudet, P., Gross-Bellard, M., and Chambon, P., *Cell* **4,** 281 (1975).
114a. Paik, W. K., and Kim, S., *Arch. Biochem. Biophys.* **134,** 632 (1969).
115. Paul, J., and More, I. A. E., *Exp. Cell Res.* **82,** 399 (1973).
116. Rill, R. I., Oosterhof, D. K., Hozier, J. C., and Nelson, D. A., *Nucleic Acids Res.* **2,** 1525 (1975).
117. Rill, R. I., and Van Holde, K. E., *J. Biol. Chem.* **248,** 1080 (1972).
118. Roberts, D. B., and Andrews, P. W., *Nucleic Acids Res.* **2,** 1291 (1975).
119. Roizes, G., *Nucleic Acids Res.* **1,** 1099 (1974).
120. Ruiz-Carrillo, A., Wangh, L. J., and Allfrey, J. G., *Science* **190,** 117 (1975).
121. Russev, G., and Tsanev, R., *Nucleic Acids Res.* **3,** 697 (1976).

122. Saffhill, R., and Itzhaki, R. F., *Nucleic Acids Res.* **2,** 113 (1975).
123. Sahasrabuddhe, C. G., and Van Holde, K. E., *J. Biol. Chem.* **249,** 152 (1974).
124. Seale, R. L., *Cell* **9,** 423 (1976).
125. Senior, M. B., Olins, A. L., and Olins, D. E., *Science* **187,** 173 (1975).
126. Severin, E. S., Kochetkov, S. M., Nesterova, M. V., and Gulyaev, N. N., *FEBS Lett.* **49,** 61 (1974).
127. Shaw, B. R., Herman, T. M., Kovacic, R. T., Beaudreau, G. S., and Van Holde, K. E., *Proc. Natl. Acad. Sci. U.S.A.* **73,** 505 (1976).
128. Shooter, K. V., Goodwin, G. H., and Johns, E. W., *Eur. J. Biochem.* **47,** 263 (1974).
129. Silver, L. M., and Elgin, S. C. R., *Proc. Natl. Acad. Sci. U.S.A.* **73,** 423 (1976).
130. Simpson, R. T., *Proc. Natl. Acad. Sci. U.S.A.* **73,** 423 (1976).
131. Simpson, R. T., and Bustin, M., *Biochemistry* **15,** 4305 (1976).
132. Simpson, R. T., and Polacow, I., *Biochem. Biophys. Res. Commun.* **55,** 1078 (1973).
133. Simpson, R. T., and Reeck, G. R. *Biochemistry* **12,** 3853 (1973).
134. Simpson, R. T., and Whitlock, J. P., Jr., *Nucleic Acids Res.* **3,** 117 (1976).
134a. Smerdon, M. J., and Isenberg, I., *Biochemistry* **15,** 4242 (1976).
135. Sobell, H. M., Tsaie, C.-C., Gilbert, S. G., Jain, S. C., and Sakore, T. D., *Proc. Natl. Acad. Sci. U.S.A.* **73,** 3068 (1976).
136. Sollner-Webb, B., Camerini Otero, R. D., and Felsenfeld, G., *Cell* **9,** 179 (1976).
137. Sollner-Webb, B., and Felsenfeld, G., *Biochemistry* **14,** 2915 (1975).
138. Sperling, J., *Photochem. Photobiol.* **23,** 323 (1976).
139. Sperling, J., and Havron, A., *Biochemistry* **15,** 1489 (1976).
140. Sperling, R., and Bustin, M., *Nucleic Acids Res.* **3,** 1263 (1976).
140a. Stedman, E., and Stedman, E., *Nature* (*London*), 556 (1943).
141. Stumph, W. F., *Endocrinology* **83,** 777 (1968).
142. Subirana, J. A., and Martinez, A. *Nucleic Acids Res.* **3,** 3025 (1976).
143. Sugano, N., and Okada, S., *Biochim. Biophys. Acta* **383,** 78 (1975).
144. Sugimura, T., Miwa, M., Kanai, Y., Oda, K., Segawa, K., Tanaka, M., and Sakura, H., *in* "Control Mechanisms in Cancer" (W. E. Criss, T. Ono, and J. R. Sabine, eds.), p. 231. Raven, New York, 1976.
144a. Sung, M. T., and Dixon, G.H., *Proc. Natl. Acad. Sci. U.S.A.* **67,** 1616 (1970).
144b. Sung, M. T., Dixon, G.H., and Smithies, O. *J. Biol. Chem.* **246,** 1358 (1971).
144c. Tabin, R., and Seligy, V., *J. Biol. Chem.* **250,** 358 (1975).
145. Thomas, J. O., and Fuber, V., *FEBS Lett.* **66,** 274 (1976).
146. Tsai, Y. H., Ansevin, A. T., and Hnilica, L. S., *Biochemistry* **14,** 1257 (1975).
147. Van Holde, K. E., and Isenberg, I., *Acc. Chem. Res.* **8,** 335 (1975).
148. Van Holde, K. E., Sahasrabuddhe, C. G., and Shaw, B. R., *Nucleic Acids Res.* **1,** 1579 (1974).
149. Van Holde, K. E., Sahasrabuddhe, C. G., Shaw, B. R., Van Bruggen, E. F. J., and Armberg, A. C. *Biochem. Biophys. Res. Commun.* **60,** 1365 (1974).
149a. Van Lente, F., Jackson, J. F. and Weintraub, H., *Cell* **5,** 45 (1975).
150. Varshavsky, A. J., Babayer, V. V., Ilyin, Y. V., Bayer, A. A., and Georgiev, G. P., *Eur. J. Biochem.* **66,** 211 (1976).
151. Weintraub, H., *Proc. Natl. Acad. Sci. U.S.A.* **72,** 1212 (1975).
152. Weintraub, H., and Groudine, M., *Science* **193,** 848 (1976).
153. Weintraub, H., Palter, K., and Van Lente, F., *Cell* **6,** 85 (1975).
154. Weintraub, H., and Van Lente, F., *Proc. Natl. Acad. Sci. U.S.A.* **71,** 4249 (1974).
155. Weintraub, H., Worcel, A., and Alberts, B., *Cell* **9,** 409 (1976).
156. Whitlock, J. P., Jr., and Simpson, R. T., *Biochemistry* **15,** 3307 (1976).
157. Wigler, M. H., and Axel, R., *Nucleic Acids Res.* **3,** 1463 (1976).

158. Wilhelm, J. A., and McCarty, K. S., *Cancer Res.* **30,** 418 (1972).
159. Williamson, R., *J. Mol. Biol.* **51,** 157 (1970).
159a. Woodcock, C., *Science* **195,** 1350 (1977).
160. Yaneva, M., and Dessev, G., *Eur. J. Biochem.* **66,** 535 (1976).
161. Yaneva, M., and Dessev, G., *Nucleic Acids Res.* **3,** 1761 (1976).
162. Yu, S. S., Li, H. J., and Shih, T. Y., *Biochemistry* **15,** 2027 (1976).
163. Yu, S. S., Li, H. J., and Shih, T. Y., *Biochemistry* **15,** 2034 (1976).

3

Histone Phosphorylation and Chromatin Structure in Synchronized Mammalian Cells

L. R. Gurley, R. A. Tobey, R. A. Walters,
C. E. Hildebrand, P. G. Hohmann, J. A. D'Anna,
S. S. Barham, and L. L. Deaven

I. INTRODUCTION

The DNA in a eukaryotic diploid cell is estimated to be nearly 2 meters in length (20), yet this DNA is packed into a nucleus only a few microns in diameter. This extraordinary feat is accomplished largely by complexing the DNA with histone proteins, which results in a folding or coiling of the DNA into the substance we call chromatin (3, 12, 15). Chromatin has many levels of organization, producing structures ranging in size from macro-

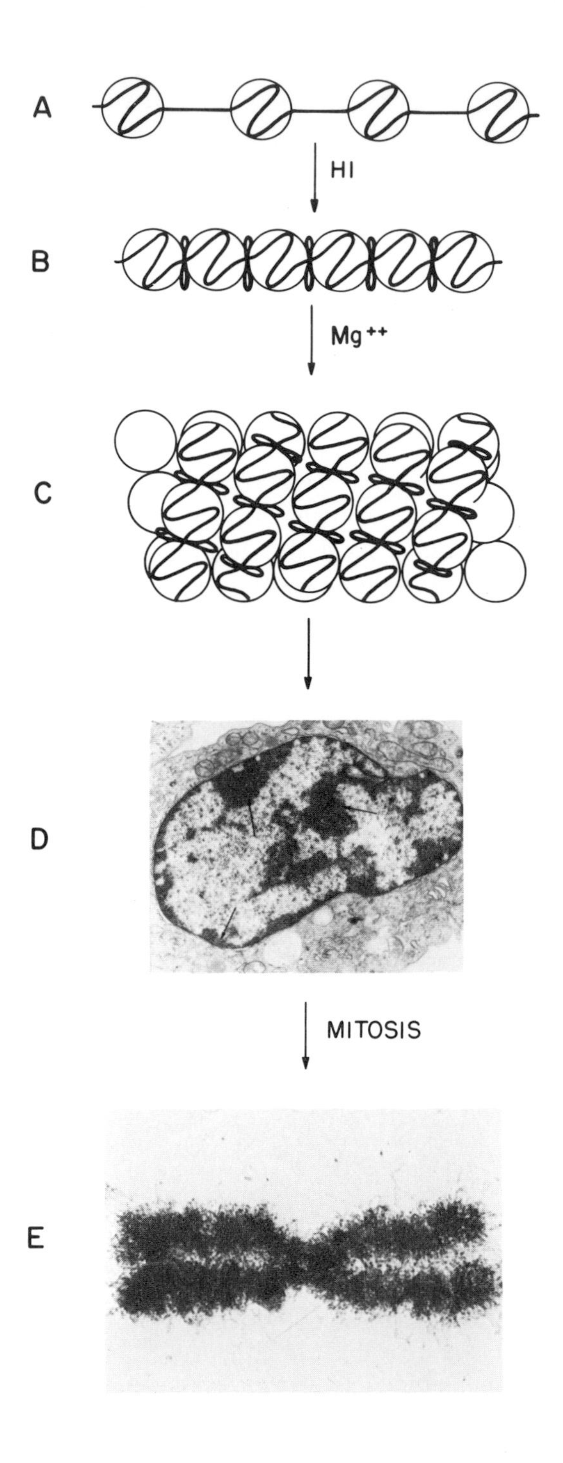
A
HI
B
Mg ++
C
D
MITOSIS
E

molecular to microscopic dimensions (69). In this hierarchy of organization, the smaller structures are incorporated into the larger structures, resulting in a reduction of the effective length of the extended DNA molecule at each level. The various orders of DNA packing (69) at each level can be described as follows.

A. First-Order Packing

According to current concepts (21), DNA is associated with repeating protein particles originally called ν bodies (56) or, more recently, nucleosomes (61). Nucleosomes are assembled from the four histones H2a, H2b, H3, and H4, but appear to be missing histone H1 (21, 45, 61). The repeating sequence of these nucleosomes in chromatin has been visualized in the electron microscope as 70-Å spheroid particles having the appearance of a "string of beads" (56, 57). When H1 is extracted from the chromatin (61) or when the chromatin fiber has been stretched out by physical forces (22, 56, 57), the nucleosomes are clearly separated from each other but remain connected by a filament of extended DNA (Fig. 1A). This stretched structure probably represents the first order of DNA packing in which a portion of the extended DNA molecule is contracted as a result of being folded, or coiled, upon the exterior of the nucleosomes (22, 64).

B. Second-Order Packing

When the chromatin fiber is not stretched or H1 is not removed, the nucleosomes exist closely packed together, forming a fairly uniform fiber approximately 100 Å in diameter called a "nucleofilament" (22) (Fig. 1B). The role of H1 in chromatin structure remains unclear, but it appears to be necessary for this second-order contraction of DNA into the nucleofilament state (22, 61). It has been proposed that H1 may interact with the strands of DNA not associated with the nucleosomes in the first-order packing state (80), causing this internucleosomal part of the DNA to be folded up (12) thus facilitating the closely packed nucleofilament structure (22).

FIG. 1. Hierarchy of chromatin organization in mammalian cells. (A) First-order packing: DNA folded on the surface of nucleosome particles and extended between these particles. (B) Second-order packing: nucleosomes closely packed together forming a fiber called a "nucleofilament." (C) Third-order packing: nucleofilament coiled into helical solenoid structure. (D) Heterochromatin: transmission electron micrograph visualizing interphase heterochromatin (arrows) thought to be aggregates of the solenoid chromatin structures. (E) Chromosome: electron micrograph of a mitotic chromosome demonstrating that it is composed of gathered fibers thought to be composed of the solenoid chromatin structure.

C. Third-Order Packing

When chromatin is prepared for electron microscopy under conditions thought to preserve its native structure, it is observed as a fiber having a diameter of 250 Å or greater (68, 69). Reasonable suggestions have been made that a higher order of DNA packing exists in this fiber, one in which the nucleofilament is coiled into a supercoil (3, 63). Recent experimental results have supported this model by demonstrating that, with the proper Mg^{2+} concentration, the nucleofilament will wind into a helical solenoid structure through the interaction of the nucleosomes (22) (Fig. 1C). This third-order packing of DNA results in a reduction of the extended DNA molecule to one-fortieth of its original length (22).

D. Higher Orders of Packing

Further condensation of the DNA occurs in both interphase and mitotic cells, but very little detail is known about its mechanism of formation or structural configuration. For example, during interphase, dense aggregates of chromatin known as heterochromatin exist (23, 69, 81) (Fig. 1D). Heitz (36) observed that these heterochromatic regions of chromatin represent portions of the chromosome which do not unravel completely during telophase (69, 81). There is evidence that these dense chromatin regions contain the supercoiled solenoid structure (17, 22, 69). Another example of higher order packing occurs during mitosis. At that time, all the chromatin fibers are folded or coiled into compact, highly condensed structures we observe in the microscope as chromosomes (Fig. 1E). Like the interphase heterochromatin, chromosomes contain the native chromatin fiber (3) which is thought to be the supercoiled solenoid structure (3, 22, 68). Thus, heterochromatin and chromosomes probably represent at least fourth-order and possibly fifth-order packings of the DNA.

For over a decade, researchers have been proposing that DNA activities (such as gene expression, gene replication, and gene segregation) might be controlled by modulating these complex molecular configurations in the chromatin structure (2, 8, 13, 32, 33, 41, 47, 50–52, 60, 74, 79). One way to accomplish such modulations without altering the DNA genetic information would be to perform reversible modifications of chromatin histones. Such modifications might be expected to result in changing the histone–DNA, histone–histone, or histone–nonhistone protein interactions that are responsible for the various chromatin structures illustrated in Fig. 1. If such a process does occur, one might expect to find correlations between histone modifications and variations in chromatin structure. We have searched for such correlations in cultured mammalian cells and have found that the modification of histones by phosphorylation is associated with chromatin condensation at a variety of levels of chromatin organization. In this chapter, we will summarize these experimental results and discuss their implications with respect to the current concepts of chromatin structure outlined above.

II. HISTONE PHOSPHORYLATION IN THE CELL CYCLE

During the course of our studies on histone metabolism in CHO Chinese hamster cells, it was observed that X-irradiation inhibited both H1 phosphorylation and cell division (28). This confirmed earlier reports by Ord and Stocken (59) and Stevely and Stocken (72), who had made similar observations in regenerating rat liver. These studies suggested that some functional relationship might exist between histone phosphorylation and cell proliferation. Support for this notion was provided by Balhorn *et al.* (6) who found a correlation between the rate of growth of a variety of tumors and the extent of H1 phosphorylation. Since the synthesis of histones is cell-cycle-dependent (27), it was felt that determining the details of histone phosphorylation with respect to the cell cycle was the next logical step in elucidating the relationship between histone phosphorylation and cell proliferation. Thus, studies on histone phosphorylation in synchronized cultures of line CHO Chinese hamster cells were initiated. The histones of cells were prelabeled with [^{3}H]lysine (30). Following synchronization by a variety of methods (76), the cells were pulse-labeled for 1 hour with $^{32}PO_4$ at various positions in their cell cycle (30–33). The histones were then isolated, fractionated, and purified by preparative electrophoresis (28). The results of these studies are summarized in Fig. 2.

Three of the five histone fractions were found to be significantly phosphorylated in proliferating CHO cells: H1, H2a, and H3 (32). Histone H4 was observed to be phosphorylated to only a limited extent (30–33). This may represent a turnover of the residual phosphate this histone retains from the cytoplasm, as described by Ruiz-Carillo *et al.* (70). No phosphorylation of H2b was observed in the chromatin of CHO cells (29).

Of the three histones significantly phosphorylated in proliferating cells, H1 was found to have the most complex cell-cycle pattern (Fig. 2A). In synchronized cells arrested in early G_1 by isoleucine deprivation, no H1 phosphorylation was observed (31). When these cells were released to traverse their cell cycle, however, H1 phosphorylation was found to occur as these cells passed through G_1 (32, 33). This $H1_{G_1}$ phosphorylation was initiated at least 2 hours before any cells entered S phase, indicating that

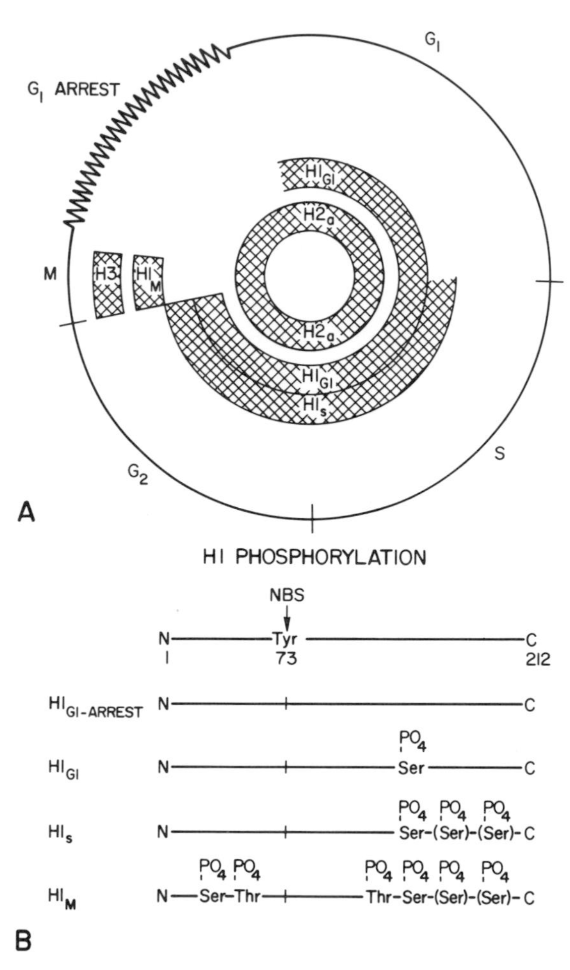

FIG. 2. Histone phosphorylation in the CHO Chinese hamster cell cycle. (A) Diagram of the cell cycle. Cells progress through the phases of this cycle clockwise from G_1 to division at the end of mitosis (M). The shaded bands indicate the phases of the cycle in which the histones (H1, H2a, and H3) are phosphorylated. (B) Diagram of the distribution of phosphorylation sites within the histone H1 molecule during the cell cycle. The regional locations of the phosphate sites in the H1 molecule were located by bisecting H1 at tyrosine residue 73 with *N*-bromosuccinimide (NBS), producing two peptide fragments that were separated by chromatography on Sephadex® G-100 (43). The amino acids phosphorylated in each fragment were then determined by high-voltage paper electrophoresis (42, 43). The number of phosphates added during the cell cycle was determined by high-resolution gel electrophoresis (43).

old histones from previous cell cycles were being phosphorylated (32–34). High-resolution gel electrophoresis by the method of Panyim and Chalkley (62) demonstrated that $H1_{G_1}$ phosphorylation involved the addition of only one phosphate to H1 (43), and column chromatography studies demonstrated that all H1 subfractions were phosphorylated in G_1 to similar extents (34).

The regional location of the $H1_{G_1}$ phosphate within the H1 molecule was determined by bisecting the molecule with *N*-bromosuccinimide (NBS) at its only tyrosine residue (43) (Fig. 2B). Analysis for phosphoserine and phosphothreonine (42) of the tryptic peptides of each part of the molecule demonstrated that $H1_{G_1}$ phosphorylation occurs on a serine in the COOH-terminal portion of the H1 (43). This type of phosphorylation was found to be cumulative, occurring on increasing numbers of H1 molecules as the cells progressed through the G_1, S, and G_2 phases (34, 43).

A second H1 phosphorylation event was observed to begin as cells entered S phase (34). This $H1_S$ phosphorylation (Fig. 2A) appears to be superimposed on the G_1 phosphorylated H1 and involves up to two additional serine sites in the COOH-terminal part of the molecule (43) (Fig. 2B). The $H1_S$ phosphorylation differs from the $H1_{G_1}$ phosphorylation in that this phosphorylated form does not accumulate as the cells traverse S and G_2 (34). Thus, only a small portion of the H1 molecules exist in the $H1_S$ phosphorylated form during late interphase (34).

A third H1 phosphorylation event was observed in CHO cells during mitosis. This $H1_M$ phosphorylation (Fig. 2A) has been described as "superphosphorylation," since it involves many phosphorylation sites and occurs on all H1 molecules (32, 34, 42, 43). Structural analysis revealed that $H1_M$ phosphorylation involves both serine and threonine sites in both NBS fragments of the molecule (43) (Fig. 2B). High resolution gel electrophoresis indicated that the major H1 subfraction contains four phosphates, while the minor H1 subfraction contains six (43). Peptide analysis demonstrated that the interphase phosphorylation sites were among the $H1_M$ phosphorylation sites (43).

During mitosis, histone H3 is also phosphorylated. This is the only time during the CHO cell cycle that this histone is observed to be phosphorylated (32, 34) (Fig. 2A). At cell division, both $H1_M$ and H3 phosphates are removed rapidly from these histones as the cells move into G_1 (32, 34).

The most rapidly phosphorylated histone in CHO cells is histone H2a (28). Unfortunately, studies on synchronized cells have not been very successful in providing insight into the function of this phosphorylation, for its phosphorylation is constitutive with respect to the cell cycle (30–33). H2a is phosphorylated at all times, both during isoleucine-deprived G_1 arrest and during all phases of the cell cycle in proliferating cells (Fig. 2A). There does appear to be a quantitative cell-cycle difference in the rate of H2a phosphorylation, however, with H2a being phosphorylated less rapidly during G_1 and most rapidly during G_2 (30, 31). These quantitative differences may reflect cell-cycle-dependent changes in chromatin structure which will be discussed later in this chapter.

Our observations of histone phosphorylation in the CHO cell cycle (Fig. 2) appear to be in general agreement with observations in other mammalian cell systems. For example, the absence of H1 phosphorylation during G_1 arrest has been reported by Oliver *et al.* (58) for stationary phase HTC cells in culture and by Thomas and Hempel (75) for stationary phase Ehrlich ascites tumor cells. Likewise, no extensive H1 phosphorylation was observed by Balhorn *et al.* (5) and Garrard *et al.* (24) in normal nonproliferating rat liver.

With respect to $H1_{G_1}$ phosphorylation in proliferating cells, Marks *et al.* (52) observed an accumulation of ^{32}P in H1 during G_1 in HeLa cells, and, more recently, Balhorn *et al.* (7) have made similar observations in HTC cells. Under more physiological conditions, Stevely and Stocken (72) observed that H1 phosphorylation preceded DNA synthesis in regenerating rat liver, and Marsh and Fitzgerald (53) made similar observations in regenerating pancreas. Garrard *et al.* (24) have recently confirmed this $H1_{G_1}$ phosphorylation in regenerating rat liver. Thus, the phosphorylation of H1 during G_1 traverse does not appear to be a peculiarity of cultured cells but, rather, is a general phenomenon of proliferating animal cells.

Like CHO, HeLa cells (7), HTC cells (7), and regenerating rat liver (24, 72) were observed to exhibit a greater H1 phosphorylation in S phase than in G_1, and Cross and Ord (16) have also observed a large increase in H1 phosphate content in PHA-stimulated lymphocytes during the DNA synthetic period.

The superphosphorylation of H1 during mitosis also appears to be a general phenomenon of mammalian cells. In addition to its occurrence in CHO cells, Lake *et al.* (46–48) have reported details of this event in Chinese hamster V-79 cells, HeLa cells, and rat nephroma cells, as has Marks *et al.* (52) and Ajiro *et al.* (1) for HeLa cells and Balhorn *et al.* (4, 7) for HTC cells and regenerating rat liver. It is now quite clear that the phosphorylation of H1 during mitosis is distinctly different from that observed in interphase. The phosphorylation of unique H1 amino acid residues during mitosis has been observed in Chinese hamster V-79 cells (49), HTC cells (7), and HeLa cells (1), as well as in CHO (42, 43), as shown in Fig. 2B. Thus, one might expect this particular H1 phosphorylation event to have a function different from the interphase H1 phosphorylation but probably common to all mammalian cells.

The specific phosphorylation of H3 during mitosis has been reported also in several different cell types: in Chinese hamster V-79 cells by Lake *et al.* (46, 47), in HTC cells by Balhorn *et al.* (7), as well as in CHO cells (30, 32, 34). The lower frequency of reports of histone H3 phosphorylation probably results from the high phosphatase activity present in mitotic cells. We found that both mitotic phosphorylated H3 and $H1_M$ were dephosphorylated during histone preparation if sodium bisulfite was not used to inhibit this phosphatase activity (34).

The constitutive nature of H2a phosphorylation with respect to the cell cycle has been noted in Ehrlich ascites tumor cells (75) and in HeLa cells (52), as well as in CHO cells (29–33). Most recently, Garrard *et al.* (24) have also demonstrated this characteristic of H2a phosphorylation in regenerating rat liver.

Because of these agreements between observations on cell-cycle-dependent histone phosphorylation in such a wide variety of cell systems, we feel our view of histone phosphorylation in the CHO cell cycle (Fig. 2) probably represents the general nature of histone phosphorylation in mammalian cells.

III. HISTONE PHOSPHORYLATION DURING MITOTIC CHROMATIN CONDENSATION

The observation of $H1_M$ and H3 phosphorylation during mitosis in mammalian cells has naturally led to suggestions that these phosphorylations may be involved in some way with the structural changes that chromatin undergoes during this phase of the cell cycle (30, 32, 34, 46, 47, 52). Similar suggestions have been made by Bradbury *et al.* (9–11) who observed extensive H1 phosphorylation during late G_2 in the slime mold *Physarum polycephalum*. These suggestions have raised questions concerning details of the relationship between histone phosphorylation and the mitotic process. For example, does mitotic histone phosphorylation begin in late G_2, or is it precisely a mitotic event? Does histone phosphorylation correlate with late G_2 chromatin condensation, with prophase chromosome organization, or only with the metaphase chromosome structural state? Are all H1 and H3 molecules phosphorylated in mitosis? Are $H1_M$ and H3 phosphorylations simultaneous, or is one histone phosphorylated before the other? Does the dephosphorylation of these histones correlate with the unraveling of chromosomes in telophase or with the dispersion of chromatin in early G_1? To answer some of these questions, histone phosphorylation was measured in synchronized cultures of CHO cells as the cells entered and exited mitosis.

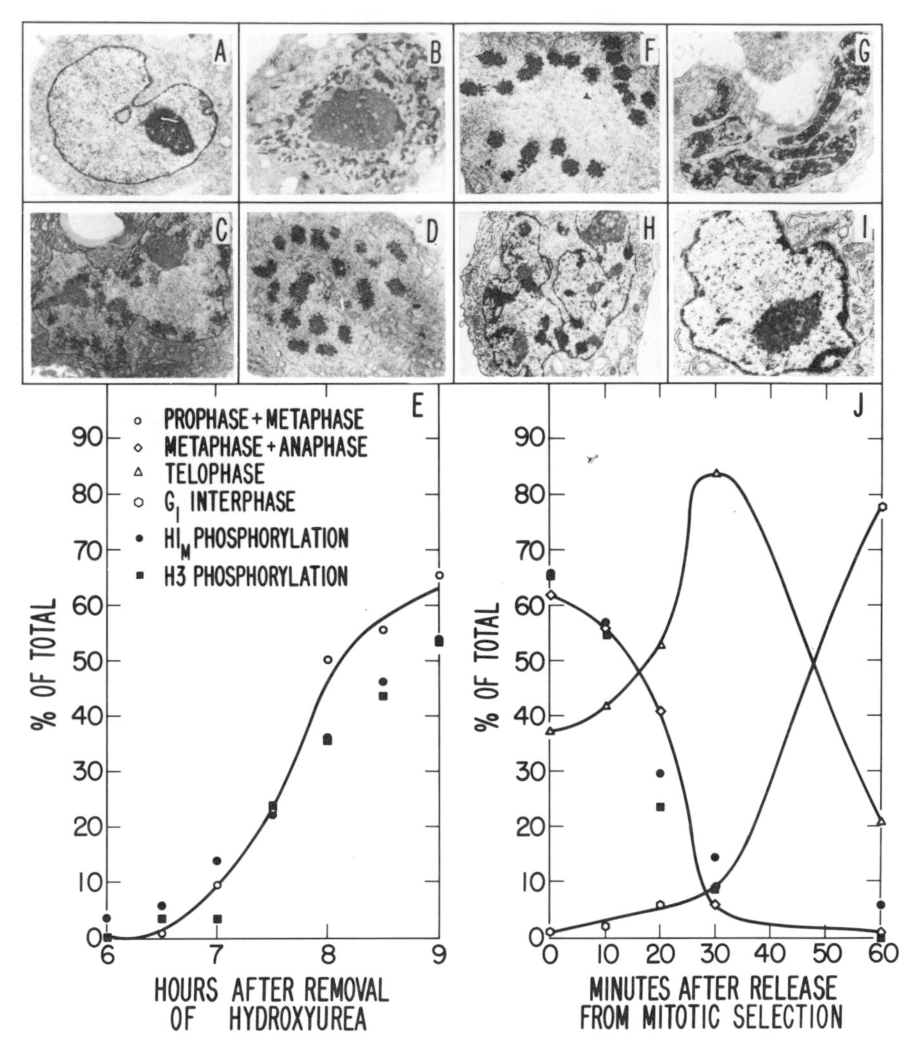

FIG. 3. Histone phosphorylation during mitosis in synchronized CHO cells. (A–E) Histone phosphorylation during chromatin condensation. Cells synchronized near the G_1/S boundary by hydroxyurea blockade were released to traverse their cell cycle to metaphase in the presence of Colcemid. The state of the chromatin structure and the phase of mitosis were determined by transmission electron microscopy at 30-minute intervals. (A) Late S phase cell. (B) Late G_2 phase cell. (C) Prophase cell. (D) Metaphase cell. (E) Percent of the H3 molecules phosphorylated (■) and percent of the H1 molecules superphosphorylated (●) increased simultaneously with the increase in percent of the cells in prophase + metaphase (○). (F–J) Histone dephosphorylation during chromosome decondensation. Cells synchronized by mitotic selection were released to traverse their cell cycle into G_1. (Note: No Colcemid was used in these cultures.) The state of the chromatin structure and the phase of mitosis were determined

Simultaneous cell samples were also taken for an examination of the chromatin structures by electron microscopy. The results of these experiments are summarized in Fig. 3.

The entry of cells into mitosis was studied after releasing synchronized cells from hydroxyurea blockade near the G_1/S boundary (76). As a result of synchrony decay, any cell population taken 7 to 9 hours after release would normally contain both early and late mitotic cells, making it impossible to distinguish differences in the biochemical processes of prophase and telophase. To prevent this mixing of early and late mitotic events, Colcemid was added to the interphase culture 4 hours after release from hydroxyurea so that cell-cycle progression would be arrested when the cells reached metaphase. Four different types of chromatin structure were observed in these cell populations: late S phase nuclei containing almost no heterochromatin (Fig. 3A), late G_2 nuclei containing large amounts of condensing chromatin (Fig. 3B), prophase nuclei containing organizing chromosomes (Fig. 3C), and metaphase cells containing fully condensed chromosomes (Fig. 3D). At various times during progression of these cultures through G_2 and into mitosis, histones were extracted and subjected to high resolution electrophoresis in long polyacrylamide gels (62). The percent of H3 molecules existing in the phosphorylated form and the percent of H1 molecules existing in the superphosphorylated $H1_M$ form (Fig. 2B) were estimated from the histone mobility shift on the polyacrylamide gels (43). The lesser phosphorylated H1 forms (i.e., three phosphates or less) associated with interphase (Fig. 2B) were not included in the $H1_M$ phosphorylation estimate. These histone estimates were then related to the percent of the cell population existing in various states of chromatin condensation shown in Fig. 3A–D. It was found that the temporal increase in $H1_M$ and H3 phosphorylation correlated with the sum of the prophase and metaphase cells existing in the population (Fig. 3E). The percent of histone molecules existing as phosphorylated H3 and superphosphorylated $H1_M$ was essentially the same during entry into mitosis, indicating that both histones were phosphorylated simultaneously.

It is concluded that these histone phosphorylations are strictly mitotic events in CHO cells, beginning with the final organization of chromatin into chromosome structures in prophase and continuing during full condensation

by transmission electron microscopy at 10-minute intervals. (F) Metaphase or anaphase cell. (G) Mid-telophase cell. (H) Late-telophase cell. (I) G_1 phase cell. (J) Percent of the H3 molecules phosphorylated (■) and percent of the H1 molecules superphosphorylated (●) decreased simultaneously with the decrease in percent of the cells in metaphase + anaphase (◊). No correlation could be observed between $H1_M$ and H3 phosphorylation and the percent of cells in telophase (△) or G_1 (⬭).

in metaphase. The preprophase gathering of chromatin into dense heterochromatinlike structures in late G_2 (Fig. 3B) does not appear to be related to these histone phosphorylation events.

The phosphorylation of histones during exit of cells from mitosis was studied, using cell populations synchronized by mitotic selection (78). This procedure produced cultures having an initial mitotic population of 99% *without* the use of Colcemid. Four different types of chromatin structures were observed in these cell populations as they moved into G_1: metaphase or anaphase cells having fully condensed chromosome structures (Fig. 3F), mid-telophase cells in which the nuclear membrane had reformed around the individual chromosomes that were beginning to unravel (Fig. 3G), late-telophase cells in which the individual nuclear membranes had coalesced to form an intact nucleus containing large amounts of its chromatin in a heterochromatin-like state (Fig. 3H), and G_1 nuclei in which the chromatin had dispersed (Fig. 3I). The percent of histone molecules existing in the $H1_M$ and H3 phosphorylation states was measured and related to the percent of the cell population existing in the various states of chromatin condensation shown in Fig. 3F–I. It was found that the temporal decrease in $H1_M$ and H3 phosphorylations correlated with the decrease in the metaphase + anaphase population, rather than with the decrease in the telophase population (Fig. 3J). There were no significant differences in the percent of histone molecules existing as phosphorylated H3 or superphosphorylated $H1_M$, and both histones were dephosphorylated simultaneously. It should be understood that lesser phosphorylated H1 forms and phosphorylated H2a still exist in telophase due to their slower dephosphorylation rate (32). Thus, the $H1_M$ and H3 dephosphorylations represent a specific removal of phosphates associated with cycle traverse out of anaphase.

It is concluded that H3 and $H1_M$ phosphorylations are associated only with the chromosome structural state per se. The dephosphorylation of these histones is immediately followed by the unraveling of chromosomes to form a dense heterochromatin-like chromatin structure in telophase. This conclusion is reinforced by the extremely good quantitative agreement between the phosphorylated histone data and the chromatin morphological data (Fig. 3J). Since mitotic selection removes telophase cells, as well as metaphase + anaphase cells, from monolayers, almost 40% of the cell population produced by mitotic selection existed in the telophase state at the beginning of the experiment. The percent of histone molecules existing in the $H1_M$ and H3 phosphorylated state at this time was the same as the percent of cells in metaphase + anaphase, the telophase cells apparently making no contribution to these phosphorylated histones. Thus, it is concluded that all histone H1 and H3 molecules are fully phosphorylated in cells containing chromosome structures. The small loss of phosphorylated

$H1_M$ and H3 in the latter stages of the experiment in Fig. 3E is attributed to dephosphorylation during histone preparation as a result of the inability of sodium bisulfite to inhibit completely the phosphatase activity that accumulates in Colcemid-blocked cells (34). However, since no Colcemid or other drugs were used in the experiment shown in Fig. 3J, there should no longer be any argument that $H1_M$ and H3 phosphorylations are artifacts of Colcemid treatment.

The absence of $H1_M$ and H3 phosphorylations in late G_2 cells or in telophase cells indicates that these phosphorylations are not involved with generation or maintenance of the dense aggregates of chromatin observed immediately before or immediately after the chromosome state. Rather, these specific phosphorylations occur only during the final organization and maintenance of chromosomes. This very close temporal and quantitative correlation offers a strong argument for the existence of a functional relationship between these specific histone phosphorylations and the final organization of chromatin (Fig. 1D) into chromosomes (Fig. 1E). These correlations do not distinguish between cause and effect, however. Thus, the questions are raised: Do the $H1_M$ and H3 phosphorylations drive the final organization of chromatin into chromosomes, or does the chromosome structural state make sites available for phosphorylation, which were inaccessible previously to kinase activity? These questions remain to be answered.

IV. HISTONE PHOSPHORYLATION DURING INTERPHASE CONDENSATION

The observations that H1 phosphorylation occurs in all cell-cycle phases of proliferating cells (Fig. 2A) suggested that interphase H1 phosphorylation might be associated with some cell growth process other than DNA synthesis. In our laboratory, Hildebrand and Tobey (37) have shown that the fraction of DNA associated with a lipoprotein complex increased in synchronized cultures following release from G_1 arrest. It was noticed that this increase was temporally correlated with the G_1 increase in $H1_{G_1}$ phosphorylation (32–34). This suggested to us that interphase H1 phosphorylation might be involved with submicroscopic changes in chromatin structure during interphase, as had been hypothesized previously (54). Changes in chromatin organization during the cell cycle had been reported by several investigators using probes of chromatin structure that interacted with the DNA component of chromatin (55, 66, 67). However, our observations on $H1_{G_1}$ phosphorylation did not appear to correlate with those measurements, especially in G_1 (38). Therefore, chromatin structure was investigated, using heparin, a different type of probe that interacts primarily with the histone component of chromatin (39, 40, 44) instead of with the DNA component. The results of these experiments are summarized in Fig. 4.

When haparin is added to nuclei, this polyanion interacts with histones, causing DNA to be released into the suspending solution (38–40). If the cells used have been grown in the presence of ^{14}C-thymidine, the released DNA can be measured easily by radioassay (38). Such measurements were made on nuclei isolated from CHO cells traversing G_1 and S phase following synchronization in G_1 arrest by isoleucine deprivation (Fig. 4A). It was found that there was an increased resistance of chromatin to heparin-mediated release of DNA as cells traversed G_1 (Fig. 4B). That is, it took progressively greater concentrations of heparin to release the same amount of DNA as the cells traversed through G_1 (38–40). These measurements were interpreted as indicating that a progressive inaccessibility of histones to heparin occurs as cells traverse the early part of interphase. The simplest model to explain these observations is that the chromatin structure becomes progressively more compact at some level of organization (Fig. 1) as cells traverse the G_1 phase.

This presumptive G_1 chromatin condensation occurred simultaneously with the $H1_{G_1}$ phosphorylation, which was initiated in G_1 following release from G_1 arrest (Fig. 4B). This correlation in cell-cycle timing suggested that some relationship exists between $H1_{G_1}$ phosphorylation and the G_1 chromatin structural changes detected with the heparin probe. This was reinforced by the fact that the sudden appearance of unphosphorylated, newly synthesized H1 in the chromatin at the beginning of S phase (Fig. 4A) resulted in a perturbation of both the H1 phosphorylation measurements and the heparin titration curve (Fig. 4B).

Similar studies were made in late interphase by resynchronizing cells near the G_1/S boundary with hydroxyurea (77). Following release from hydroxyurea blockade (Fig. 4C), the resistance of chromatin to heparin-mediated release of DNA continued to increase progressively until cell division, at which time the resistance to the heparin probe returned to low values as the cells entered G_1 (Fig. 4D). During S and G_2 traverse, the H1 phosphorylation, which is composed primarily of the $H1_{G_1}$ type (34) also increased, suggesting that the relationship between these two measurements extends across the whole interphase process. H1 phosphorylation measurements were not taken after 6 hours in Fig. 4D due to the initiation of $H1_M$ superphosphorylation associated with cell division.

Mazia (54) has suggested that the changes in chromatin structure that can be seen from prophase to telophase (Fig. 3) might be only part of a greater chromosome structure cycle that extends from G_1 to mitosis, most of which is not visible at the microscopic level. Pederson (66) has further

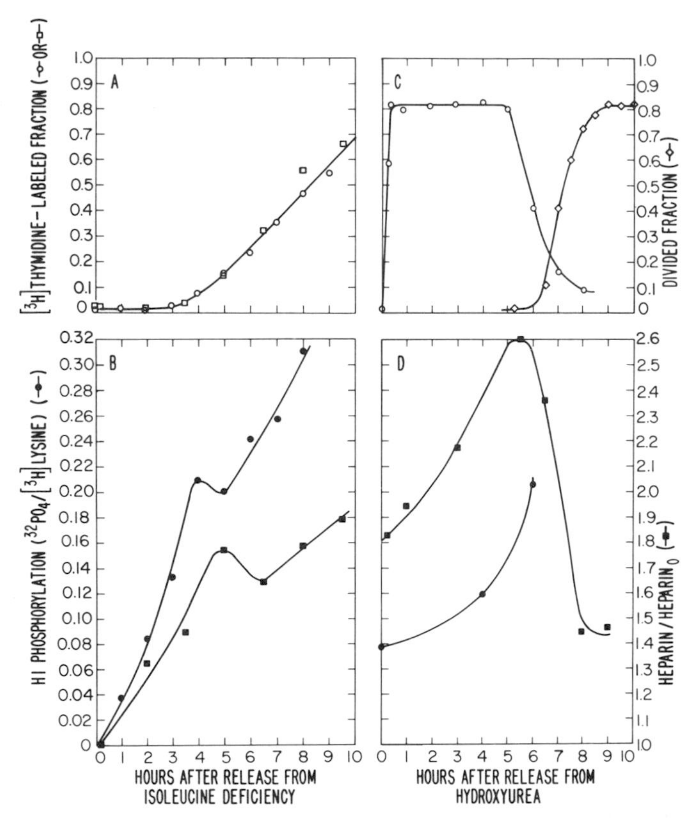

FIG. 4. Cell-cycle-dependent histone H1 phosphorylation changes related to chromatin structural organizational changes detected by heparin titration of CHO nuclei. (A) Cells were synchronized in G_1 arrest by isoleucine deprivation and then released to traverse G_1 and S phases. Entry into S phase was detected by incorporation of [^{3}H]thymidine into DNA [(○) or (□), two different cultures]. (B) The increase in rate of interphase H1 phosphorylation detected by $^{32}PO_4$ incorporation (●) is compared to the increase in resistance of chromatin to disruption by heparin treatment (■) during G_1 and S phase traverse. (C) Cells were resynchronized near the G_1/S boundary by hydroxyurea treatment and then released to traverse S and G_2 phases. Entry into and exit from S phase were detected by incorporation of [^{3}H]thymidine into DNA (○). Synchronous cell division was monitored by counting culture cell density (◊). (D) The increase in rate of interphase H1 phosphorylation (●) is compared to the increase in resistance of chromatin to disruption by heparin treatment (■) during S and G_2 phase traverse. The increases in resistance of chromatin to disruption by heparin treatment in (B) and (D) are expressed as increases in the ratio of the concentration of heparin required to release 50% of the dissociatable DNA at any time in the cell cycle [heparin], to the same measurement made at the beginning of the cell cycle [heparin]$_0$.

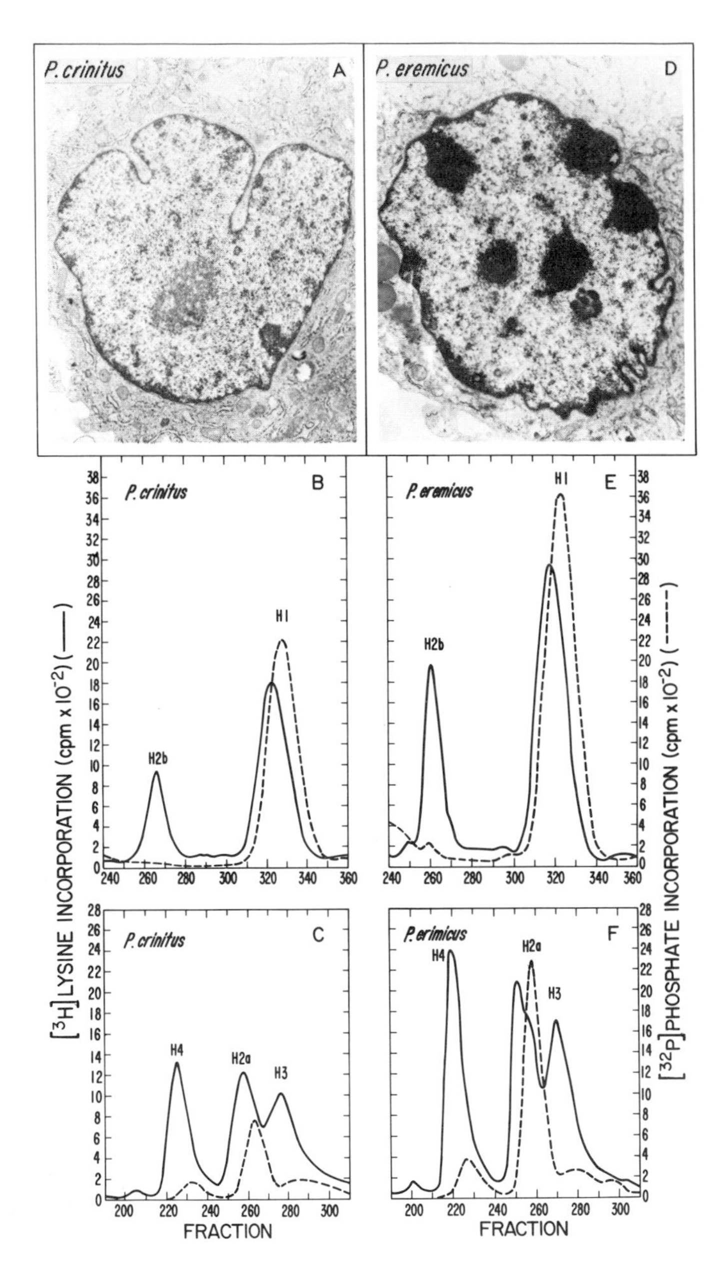

FIG. 5. Histone phosphorylation in two *Peromyscus* cell lines having different quantities of constitutive heterochromatin. (A) Transmission electron microscopy of an interphase *P. crinitus* cell demonstrating its low heterochromatin content. (B and C) Preparative electrophoresis of histones from *P. crinitus* cells labeled 2.3 generations with [^{3}H] lysine and then

suggested that this chromosome cycle might be interrupted in nondividing cells. He points out that reinitiation of the chromatin structural changes of the chromosome cycle might lead to the activation of those biochemical events of the cell cycle which lead to cell proliferation. In addition, Stubblefield (73) has made cytological observations that support the idea that a preparatory process (presumably chromatin structural changes) occurs in G_2, which is necessary for proper chromosome formation in mitosis. We feel that the interphase chromatin structural changes detected by the heparin probe (Fig. 4) support the concept of an interphase chromatin structural cycle and that the close correlation of $H1_{G_1}$ phosphorylation with these structural changes implicates an involvement of this histone phosphorylation in the process.

As is the case with mitotic histone phosphorylations, the correlations between $H1_{G_1}$ phosphorylation and presumptive interphase chromatin condensation detected by the heparin probe do not distinguish between cause and effect. Thus, again the questions are raised: Does $H1_{G_1}$ phosphorylation drive interphase chromatin structural changes, or do the changes in chromatin structure make sites available for H1 phosphorylation?

V. HISTONE PHOSPHORYLATION AND HETEROCHROMATIN

As outlined above, our experience suggested that histone phosphorylation may be associated with condensing chromatin structures. To determine whether or not this association might represent a general phenomenon, we attempted to obtain evidence for such a correlation in another cell system. Interesting results were obtained when we compared histone phosphorylation in two cell lines of deer mice, *Peromyscus eremicus* and *Peromyscus crinitus,* which possess similar levels of euchromatin but which differ greatly in their contents of constitutive heterochromatin (65). In particular, we had anticipated that we might find a higher level of H1 phosphorylation in the high heterochromatin-containing *P. eremicus* cells than in the low heterochromatin-containing *P. crinitus* cells. To our surprise, however, we found a correlation between histone H2a phosphorylation and the amount of heterochromatin. The results of these experiments are summarized in Fig. 5.

pulse-labeled with $^{32}PO_4$ for 2 hours during exponential growth. (B) $^{32}P/^{3}H$ ratio of H1 was 1.203. (C) $^{32}P/^{3}H$ ratio of H2a was 0.494. (D) Transmission electron microscopy of an interphase *P. eremicus* cell demonstrating its excessive heterochromatin content. (E and F) Preparative electrophoresis of histones from *P. eremicus* cells labeled 2.3 generations with [^{3}H]-lysine and then pulse-labeled with $^{32}PO_4$ for 2 hours during exponential growth. (E) $^{32}P/^{3}H$ ratio of H1 was 1.200. (F) $^{32}P/^{3}H$ ratio of H2a was 0.779.

Cell lines from ear cultures of these two closely related species of deer mice were obtained as a generous gift from Dr. T. C. Hsu of the Department of Biology, M. D. Anderson Hospital and Tumor Institute of the University of Texas at Houston. Both cell lines had a diploid chromosome number of 48 (65). G-banding analysis demonstrated that the euchromatin contents of these two species were virtually the same (65). Cytologically, the difference between these two cell lines lay in their short chromosome arms. Most of the chromosomes of *P. eremicus* cells contained short arms composed almost entirely of constitutive heterochromatin, while the chromosomes of *P. crinitus* cells were missing these heterochromatic short arms (65). From flow microfluorometric measurements, it was shown that *P. eremicus* cells contained 36% more DNA than did *P. crinitus* cells (35), in excellent agreement with chromosome arm length measurements (65). This indicated that the extra DNA of *P. eremicus* cells existed as highly condensed constitutive heterochromatin in the short chromosome arms. Both light and electron microscopy demonstrated that the extra heterochromatin observed in metaphase chromosomes of *P. eremicus* cells persisted in interphase as large, dense clumps of dark-staining heterochromatin (35, 65) (Fig. 5D). Interphase *P. crinitus* cells contained only small clumps of heterochromatin (Fig. 5A).

To determine if histone phosphorylation is associated with the degree of interphase chromatin condensation, we compared the extent of incorporation of $^{32}PO_4$ into the histones of these two cell lines growing under identical culture conditions (35). Cells were grown in the presence of ^{3}H-lysine for exactly the same number of cell divisions to label equally their histones. The cells were then pulse-labeled with $^{32}PO_4$ for 2 hours in exponential growth. Following histone isolation, the lysine-rich histones (Figs. 5B and 5E) and the arginine-rich histones (Figs. 5C and 5F) were fractionated and purified by preparative electrophoresis. The *P. eremicus* cells contained more histone (Figs. 5E and 5F) than did the *P. crinitus* cells (Figs. 5B and 5C), as expected from the relative DNA contents of these cells, but both cell lines had essentially the same relative proportion of the five histones. The relative rates of histone phosphorylation were determined from the $^{32}P/^{3}H$ ratio in the electrophoretic peaks (35). Surprisingly, the H1 phosphorylation rate in *P. crinitus* cells (Fig. 5B) was identical to the H1 phosphorylation rate in *P. eremicus* cells (Fig. 5E). However, in *P. eremicus* cells, which have the greater amount of interphase heterochromatin, the H2a phosphorylation rate was 58% greater (Fig. 5F) than in *P. crinitus* cells, which have a small amount of heterochromatin (Fig. 5C).

We conclude that H1 phosphorylation is not likely involved with the processes which form or maintain the visible heterochromatin aggregates during interphase (Fig. 1D). On the other hand, these experiments do suggest that H2a phosphorylation may be involved with interphase heterochromatin structure in some way. The lack of cell-cycle specificity of H2a phosphorylation (Fig. 2A) might be expected if this phosphorylation were involved with such a general interphase chromatin structure as heterochromatin.

While H2a phosphorylation occurs at all times during the cell cycle, the rate of this phosphorylation increases as cells traverse the cell cycle, being 1.63 times greater in G_2-rich cultures than in mid-S phase cultures (30, 31). If H2a phosphorylation is involved in heterochromatin structure, one might expect an increase in the relative proportion of chromatin existing as heterochromatin as cells move from S to G_2. Using microdensitometry, de la Torre and Navarrete (18) and Sawicki *et al.* (71) have measured the amount of condensed chromatin at different stages of interphase. They found that there was an increase in the proportion of dense chromatin during traverse from S to G_2, and de la Torre *et al.* (19) have estimated that the proportion of dense chromatin is 1.85 times greater in G_2 cells than in S-phase cells. This increase is consistent with the increased H2a phosphorylation rates we observed during this period (1.63) of the CHO cell cycle (30).

VI. DISCUSSION

Our experience has been that, whenever we observe histones being phosphorylated, this process appears to be associated in some way with condensing or condensed chromatin structures. These observations raise the question: Is there a functional connection between histone phosphorylation and chromatin condensation? If so, by what mechanism does it work? And, is this phosphorylation the driving force for condensation, or is phosphorylation involved in a more passive way?

It should be emphasized that the experiments presented above demonstrate temporal correlations only and do not distinguish between cause and effect relationships, nor do they eliminate unrelated coincidences. Their usefulness lies in providing a sufficiently detailed description of these processes in living cells to provide a basis for theoretical considerations that can lead to the design of testable hypotheses.

Consider the case of interphase H1 phosphorylation. It is thought that H1 is involved in the formation or stabilization of the nucleofilament (Fig. 1B) and that, with the help of Mg^{2+}, the nucleosomes can then interact to coil into a more condensed solenoid structure (Fig. 1C). We had suspected initially that the addition of negatively charged phosphates to positively charged histones might weaken the interaction between DNA and histone, thus loosening the chromatin structure and allowing its decondensation.

Our observations in Fig. 4 suggest that, instead, perhaps we should be asking how could interphase H1 phosphorylation assist in the formation of a more stable condensed structure.

Several possible mechanisms are worthy of consideration. Phosphorylation of H1 might produce a twist to the nucleofilament, which favors coiling. In a more passive role, the phosphorylation of H1 might remove some structural constraint that the unphosphorylated H1 might exert on the nucleofilament to prevent coiling. As an example of a highly passive role, coiling of the nucleofilament might be driven by a totally independent process, exposing some sites of H1 to phosphorylation that were inaccessible in the extended nucleofilament form. In this case, the formation of H1 phosphate–magnesium complexes with nuclear proteins or with DNA might stabilize the solenoid structure. These speculative questions and suggested mechanisms illustrate that, by combining our recently acquired knowledge of chromatin structure and histone phosphorylation, we can now ask more specific and interesting questions concerning the function of histone phosphorylation.

As mentioned earlier, our cell-cycle studies with CHO cells had not given us much insight into the function of constitutive H2a phosphorylation. However, the results obtained with the two unique mouse cell lines shown in Fig. 5 have suggested a role of H2a phosphorylation in heterochromatin structure. Since it is thought that heterochromatin (Fig. 1D) is composed primarily of chromatin packaged in the solenoid form (Fig. 1C), how might one envision a role for H2a phosphorylation in this chromatin aggregate? Since H2a is a part of the nucleosome, its phosphorylation might lead to bridging between the nucleosomes of adjacent solenoids through Mg^{2+} or Ca^{2+} complexes or through complexes with bridging proteins. Comings and Harris (14) recently have reported finding some unique nonhistone proteins in heterochromatin that might serve this purpose. Our observations suggest that this possibility should be investigated.

The best correlation between histone phosphorylation and chromatin condensation is found at mitosis. The results in Fig. 3 leave no doubt that, in CHO cells, $H1_M$ and H3 phosphorylations occur precisely at the time that the chromatin fiber is folded into its compact chromosome structure (Fig. 1E). Such a close correlation offers a strong argument for the existence of some functional relationship between these two phenomena. An opposing argument can be made, however, that these mitotic-specific phosphorylations might be trivial, occurring coincidentally with chromosome condensation as a result of sudden exposure to cytoplasmic enzymes when the nuclear membrane breaks up at mitosis. Since we find these phosphorylations occurring during prophase while the nuclear membrane is still intact (Fig. 3C), we are reluctant to accept this argument.

Our observation of a precise correlation between $H1_M$ phosphorylation and chromosome condensation (Fig. 3) would appear, on casual examination, to be at variance with the reports of Bradbury *et al.* (9–11). Using *Physarum polycephalum,* these workers have made measurements demonstrating that (a) H1 phosphate content rises in the G_2 phase, reaching a peak at the beginning of early prophase, and (b) then drops "dramatically" as cells progress to metaphase (9). They have argued that these observations suggest H1 phosphorylation must be completed "before" chromosome condensation occurs (9) and that the H1 phosphorylating activity of the cell constitutes the "mitotic trigger" [i.e., the driving force for chromosome condensation (10, 11)]. We suspect those measurements differ from ours (Fig. 3) for two reasons. (a) They do not separate interphase type H1 phosphorylation from mitotic type H1 phosphorylation; thus, it seems likely they are describing the interphase type H1 phosphorylation in G_2 (see Fig. 2), while we are concerned with a different type of $H1_M$ phosphorylation in our correlations in Fig. 3. (b) Apparently, phosphatase inhibitors (such as sodium bisulfite) were not used in preparing histones during mitosis in *Physarum*. We have found that in CHO cells, omission of sodium bisulfite results in an almost complete loss of the $H1_M$ and H3 phosphates during preparation of histones from mitotic cells (32, 34). Thus, we suspect the observation of a loss of H1 phosphates in *Physarum* at precisely the time we observe maximum phosphorylation in CHO cells results from this mitotic-specific phosphatase activity dephosphorylating H1 during preparation.

The point of this duscussion is that a comparison of the measurements made in these two different cell systems should be made with caution. We observed a doubling of interphase type H1 phosphorylation during G_2 in CHO cells (30), similar to that observed in *Physarum* (9), but this is not the $H1_M$ phosphorylation in CHO. It is the type of H1 phosphorylation we find correlated with chromatin structural changes in interphase as detected by the heparin probe. This interphase H1 phosphorylation may prepare the chromatin for proper mitotic chromosome condensation, but we question whether it is the driving force for condensation, since it begins in G_1 phase in mammalian cells. If histone phosphorylation is the mitotic trigger, it seems more reasonable to us that the $H1_M$ and H3 phosphorylations are the more likely candidates for this function than the interphase type H1 phosphorylation, since $H1_M$ and H3 phosphorylations are so precisely correlated with chromosome condensation and maintenance (i.e., prophase, metaphase, and anaphase). While acknowledging that our temporal correlations provide no evidence to support the mitotic trigger hypothesis, neither do they exclude the possibility that $H1_M$ and H3 phosphorylations drive chromosome condensation during prophase. Gorovsky and Keevert (25, 26) have opposed this idea, however, arguing that, since they cannot find any

H1 or H3 histones in the mitotically dividing micronuclei of *Tetrahymena pyriformis,* the phosphorylation of these proteins cannot be the driving force for chromosome condensation.

If the Gorovsky and Keevert position (25, 26) is ultimately confirmed, is it possible to envision a role for $H1_M$ and H3 phosphorylations in chromosome condensation which is not trivial? We think so. For example, if H1 and H3 exert some structural constraint on the chromatin complex that prevents its condensation into chromosomes, phosphorylation of these proteins might unblock these restrictions, allowing chromosomes to organize at mitosis. If no H1 and H3 exist in the micronucleus of *Tetrahymena* in the first place, these restrictions would be absent in this specialized situation. In such an instance, $H1_M$ and H3 phosphorylations would not be a driving force for chromosome condensation but, nevertheless, would be a necessary process for chromosome condensation and maintenance in mammalian cells.

At present, we think that the close correlation between the specific $H1_M$ and H3 phosphorylations and chromosome condensation justifies continued consideration of this histone modification as an important process in the mitotic state. It is difficult to accept that such a generally occurring process is trivial. It is hoped that, by combining the exciting new concepts of chromatin structure with continuing developments in our knowledge of histone phosphorylation, we can propose more specific hypotheses for experimental investigation.

ACKNOWLEDGMENT

This work was performed under the auspices of the U.S. Energy Research and Development Administration.

REFERENCES

1. Ajiro, K., Borun, T. W., and Cohen, L. H., *Fed. Proc., Fed. Am. Soc. Exp. Biol.* **34,** 581 (1975).
2. Allfrey, V. G., *in* "Histones and Nucleohistones" (D. M. P. Phillips, ed.), p. 241. Plenum, New York, 1971.
3. Bahr, G. F., *Fed. Proc., Fed. Am. Soc. Exp. Biol.* **34,** 2209 (1975).
4. Balhorn, R., Rieke, W. O., and Chalkley, R., *Biochemistry* **10,** 3952 (1971).
5. Balhorn, R., Chalkley, R., and Granner, D., *Biochemistry* **11,** 1094 (1972).
6. Balhorn, R., Balhorn, M., Morris, H. P., and Chalkley, R., *Cancer Res.* **32,** 1775 (1972).
7. Balhorn, R., Jackson, V., Granner, D., and Chalkley, R., *Biochemistry* **14,** 2504 (1975).
8. Bradbury, E. M., and Crane-Robinson, C. *in* "Histones and Nucleohistones" (D. M. P. Phillips, ed.), p. 85. Plenum, New York, (1971).
9. Bradbury, E. M., Inglis, R. J., Matthews, H. R., and Sarner, N., *Eur. J. Biochem.* **33,** 131 (1973).

10. Bradbury, E. M., Inglis, R. J., and Matthews, H. R., *Nature* (*London*) **247,** 257 (1974).
11. Bradbury, E. M., Inglis, R. J., Matthews, H. R., and Langan, T. A., *Nature* (*London*) **249,** 553 (1974).
12. Bram, S. *Biochimie* **57,** 1301 (1975).
13. Chalkley, R., Balhorn, R., Oliver, D., and Granner, D., *in* "Protein Phosphorylation in Control Mechanisms" (F. Huijing and E. Y. C. Lee, eds.), p. 251–277, Vol. 5, Miami Winter Symposia. Academic Press, New York, 1973.
14. Comings, D. E., and Harris, D. C., *Exp. Cell Res.* **96,** 161 (1975).
15. Crick, F. H. C., and Klug, A., *Nature* (*London*) **255,** 530 (1975).
16. Cross, M. E., and Ord, M. G., *Biochem. J.* **118,** 191 (1970).
17. Davies, H. G., and Haynes, M. E., *J. Cell Sci.* **21,** 315 (1976).
18. de la Torre, C., and Navarrete, M. H., *Exp. Cell Res.* **88,** 171 (1974).
19. de la Torre, C., Sacristán-Gárate, A., and Navarrete, M. H., *Chromosoma* **51,** 183 (1975).
20. DuPraw, E. J., "DNA and Chromosomes," p. 118. Holt, Rinehart and Winston, Inc., New York, 1970.
21. Elgin, S. C. R., and Weintraub, H., *Annu. Rev. Biochem.* **44,** 725 (1975).
22. Finch, J. T., and Klug, A., *Proc. Natl. Acad. Sci. U.S.A.* **73,** 1897 (1976).
23. Frenster, J. H., *in* "The Cell Nucleus" (H. Busch, ed.), Vol. 1, p. 565. Academic Press, New York, 1974.
24. Garrard, W. T., Kidd, G. H., and Bonner, J. *Biochem. Biophys. Res. Commun.* **70,** 1219 (1976).
25. Gorovsky, M. A., and Keevert, J. B., *Proc. Natl. Acad. Sci. U.S.A.* **72,** 2672 (1975).
26. Gorovsky, M. A., and Keevert, J. B., *Proc. Natl. Acad. Sci. U.S.A.* **72,** 3536 (1975).
27. Gurley, L. R., and Hardin, J. M., *Arch. Biochem. Biophys.* **128,** 285 (1968).
28. Gurley, L. R., and Walters, R. A., *Biochemistry* **10,** 1588 (1971).
29. Gurley, L. R., and Walters, R. A., *Biochem. Biophys. Res. Commun.* **55,** 697 (1973).
30. Gurley, L. R., Walters, R. A., and Tobey, R. A., *Biochem. Biophys. Res. Commun.* **50,** 744 (1973).
31. Gurley, L. R., Walters, R. A., and Tobey, R. A., *Arch. Biochem. Biophys.* **154,** 212 (1973).
32. Gurley, L. R., Walters, R. A., and Tobey, R. A., *J. Cell Biol.* **60,** 356 (1974).
33. Gurley, L. R., Walters, R. A., and Tobey, R. A., *Arch. Biochem. Biophys.* **164,** 469 (1974).
34. Gurley, L. R., Walters, R. A., and Tobey, R. A., *J. Biol. Chem.* **250,** 3936 (1975).
35. Gurley, L. R., Walters, R. A., Barham, S. S., and Deaven, L. L., *Exp. Cell Res.*, in press (1977).
36. Heitz, E., *Jahrb. Wiss. Bot.* **69,** 762 (1928).
37. Hildebrand, C. E., and Tobey, R. A., *Biochim. Biophys. Acta* **331,** 165 (1973).
38. Hildebrand, C. E., and Tobey, R. A., *Biochem. Biophys. Res. Commun.* **63,** 134 (1975).
39. Hildebrand, C. E., Gurley, L. R., Tobey, R. A., and Walters, R. A., *Fed. Proc., Fed. Am. Soc. Exp. Biol.* **34,** 581 (1975).
40. Hildebrand, C. E., Walters, R. A., Tobey, R. A., and Gurley, L. R., *Biophys. J.* **16,** 226a (1976).
41. Hnilica, C. S., "The Structure and Function of Histones," p. 79. CRC Press, Cleveland, Ohio, (1972).
42. Hohmann, P.G., Tobey, R. A., and Gurley, L. R., *Biochem. Biophys. Res. Commun.* **63,** 126 (1975).
43. Hohmann, P.G., Tobey, R. A., and Gurley, L. R., *J. Biol. Chem.* **251,** 3685 (1976).
44. Kitzis, A., Defer, N., Dastague, B., Sabatier, M. M., and Kruh, J., *FEBS Lett.* **66,** 336 (1976).

45. Kornberg, R. D., *Science* **184,** 868 (1974).
46. Lake, R. S., and Salzman, N. P., *Biochemistry* **11,** 4817 (1972).
47. Lake, R. S., Goidl, J. A., and Salzman, N. P., *Exp. Cell Res.* **73,** 113 (1972).
48. Lake, R. S., *Nature* (*London*), *New Biol.* **242,** 145 (1973).
49. Lake, R. S., *J. Cell Biol.* **58,** 317 (1973).
50. Langan, T. A., *in* "Role of Cyclic AMP in Cell Function" Advances in Biochemical Psychopharmacology (P. Greengard and E. Costa, eds.), pp. 307–323, Vol. 3. Raven Press, New York. 1970.
51. Louie, A. J., Candido, E. P. M., and Dixon, G. H., *Cold Spring Harbor Symp. Quant. Biol.* **38,** 803 (1974).
52. Marks, D. B., Paik, W. K., and Borun, T. W., *J. Biol. Chem.* **248,** 5660 (1973).
53. Marsh, W. H., and Fitzgerald, J., *Fed. Proc., Fed. Am. Soc. Exp. Biol.* **32,** 2119 (1973).
54. Mazia, D. *J. Cell. Comp. Physiol.* **62,** Suppl. 1, 123 (1963).
55. Nicolini, C., Ajiro, K., Borun, T. W., and Baserga, R., *J. Biol. Chem.* **250,** 3381 (1975).
56. Olins, A. L., and Olins, D. E., *Science* **183,** 330 (1974).
57. Olins, A. L., Carlson, R. D., and Olins, D. E., *J. Cell Biol.* **64,** 528 (1975).
58. Oliver, D., Balhorn, R., Granner, D., and Chalkley, R., *Biochemistry* **11,** 3921 (1972).
59. Ord, M. G., and Stocken, L. A., *Biochem. J.* **107,** 403 (1968).
60. Ord, M. G., and Stocken, L. A., *Biochem. J.* **112,** 81 (1969).
61. Oudet, P., Gross-Bellard, M., and Chambon, P., *Cell* **4,** 281 (1975).
62. Panyim, S., and Chalkley, R., *Arch. Biochem. Biophys.* **130,** 337 (1969).
63. Pardon, J. F., and Wilkins, M. H. F., *J. Mol. Biol.* **68,** 115 (1972).
64. Pardon, J. F., Worcester, D. L., Wooley, J. C., Tatchell, K., Van Holde, K. E., and Richards, B. M., *Nucleic Acids Res.* **2,** 2163 (1975).
65. Pathak, S., Hsu, T. C., and Arrighi, F. E., *Cytogenet. Cell Genet.* **12,** 315 (1973).
66. Pederson, T., *Proc. Natl. Acad. Sci. U.S.A.* **69,** 2224 (1972).
67. Pederson, T., and Robbins, E., *J. Cell Biol.* **55,** 322 (1972).
68. Ris, H., and Kubai, D. F., *Annu. Rev. Genet.* **4,** 263 (1970).
69. Ris, H., *in* "The Structure and Function of Chromatin" (D. W. Fitzsimons and G. E. W. Wolstenholme, eds.), pp. 7–28, Vol. 28, Ciba Found. Symp. Associated Scientific Publishers, Amsterdam, 1975.
70. Ruiz-Carillo, A., Wangh, L. J., and Allfrey, V. G., *Science* **190,** 117 (1975).
71. Sawicki, W., Rowinski, J., and Swenson, R., *J. Cell. Physiol.* **84,** 423 (1974).
72. Stevely, W. S., and Stocken, L. A., *Biochem. J.* **110,** 187 (1968).
73. Stubblefield, E., *In* "Cytogenetics of Cells in Culture" (R. J. C. Harris, ed.) pp. 223–248, Vol. 3. Academic Press, New York, 1964.
74. Swift, H., *Cold Spring Harbor Symp. Quant. Biol.* **38,** 963 (1974).
75. Thomas, G., and Hempel, K., *Exp. Cell Res.* **100,** 309 (1976).
76. Tobey, R. A., *in* "Methods in Cell Biology" (D. M. Prescott, ed.), pp. 67–112, Vol. 6, Academic Press, New York, 1973.
77. Tobey, R. A., and Crissman, H. A., *Exp. Cell Res.* **75**, 460 (1972).
78. Tobey, R. A., Anderson, E. C., and Petersen, D. F., *J. Cell. Physiol.* **70,** 63 (1967).
79. Tobey, R. A., Gurley, L. R., Hildebrand, C. E., Ratliff, R. L., and Walters, R. A., *in* "Control of Proliferation in Animal Cells" (B. Clarkson and R. Baserga, eds.), Vol. 1, p. 665. Cold Spring Harbor Lab., Cold Spring Harbor, New York, 1974.
80. Van Holde, K. E., and Isenberg, I., *Acc. Chem. Res.* **8,** 327 (1975).
81. Yunis, J. J., and Yasmineh, W. G. *in* "Advances in Cell and Molecular Biology" (E. J. DuPraw, ed.), Vol. 2, pp. 1–46. Academic Press, New York, 1972.

4

The Binding of Histones in Mammalian Chromatin: Cell-Cycle-Induced and SV40-Induced Changes

Margarida O. Krause

I. INTRODUCTION

The role of histones in the structural and genetic control of chromatin is still poorly understood in spite of the relative abundance of detailed chemical knowledge of these basic nuclear proteins. Recent studies on the structure of chromatin revealed that histones associate with DNA in a regular repetitive beadlike structural unit about 10 nm in diameter, whereby approximately 140 nucleotide paris of DNA associate with a histone octamer composed of two H3–H4 dimers and a tetramer of 2H2b–2H2a molecules (1–6). These units, named nucleosomes (7) are separated by 40–

50 nucleotides and are postulated to consist of protein cores around which is wound the DNA double helix (8). Since nucleosomes appear in chromatin fibers under the electron microscope whenever preparative procedures lead to loss of H1 and divalent cations (7), it is presumed that H1 acts by forming bridges between these structural units binding to DNA at the nucleotide sequences between nucleosomes, such as to keep them in a supercoiled or solenoidal arrangement corresponding to the more commonly observed 20–30 nm thick fiber (8–10). This fiber, named "native fiber" by Ris is tightly packed in heterochromatin and loosely packed in euchromatin. A higher order of supercoiling of this native fiber gives rise to the 100-nm thick regions of interphase fibers called elementary chromomeres, or to the 100-nm thick chromonema fiber, which in turn is further folded into chromatids in metaphase chromosomes.

The uniformity of histone composition and nucleosome appearance in the chromatin fibers, regardless of DNA primary sequence or genetic activity (11–13), suggests that histones play a purely structural role and, therefore, do not participate in genetic control. Yet there are indications of nucleosome heterogeneity (14) as well as changes in histone modification reactions during different phases of gene activity which suggest that histones are also likely to be involved in modulation of gene expression (15). Moreover, studies in our laboratory indicate that the type and strength of binding of different histone fractions in chromatin is not the same in cycling and noncycling cells, as well as in normal and SV40-transformed populations.

While studying synthesis, phophorylation, and acetylation of histones in cultured mouse L cells at different phases of growth, we noticed that the relative amounts of each histone fraction recovered from either G_1-arrested (G_0) or log-phase cell populations was not the same (16). Since we thought it unlikely that cells at different phases of growth would have different relative amounts of histones, we suspected that selective leakage of some histone fractions might be occurring during cell lysis and isolation of nuclei. This selective leakage could be explained if one assumed that some histone fractions are more tightly bound in chromatin during one phase of growth than during another. We, therefore, decided to investigate this problem by trying a number of different pH and ionic conditions for isolation of nuclei. By taking advantage of conditions that elicit partial leakage of one or another histone fraction, we could then approach the problem of histone binding in chromatin in relation to gene activation. For this purpose, we studied two types of gene activation processes in cultured mammalian cells: (a) serum activation of G_1 arrested cells, and (b) transformation by SV40 virus.

II. CHANGES INDUCED BY SERUM STIMULATION OF G_0 CELLS

These studies were carried out using both mouse L cells (strain 929E) and WI-38 human diploid fibroblasts in monolayer culture. Both cell types become arrested in a G_0 stage when attaining confluency, approximately 7 days after subcultivation.

A. Lysine-Rich Histones: Extractability, Synthesis, and Phosphorylation

The effect of slight changes in the pH of the lysing medium on the retention of H1 histone from exponential and stationary (G_0) L cell nuclei is presented in Table I. When cells are lysed at pH 2.85, most, if not all, of the H1 is retained in both exponential and stationary cell nuclei. However, a decrease in only one-tenth of a pH unit results in leakage of 60% of the H1 fraction from exponential cell nuclei, with only a 20% leakage from stationary ones. Another decrease in one-tenth of a pH unit eliminates the cell cycle differences and results in substantial leakage in both cases (16). It appears, therefore, that there is a very narrow pH range within which a sizable amount of lysine-rich histone leaks out of the nuclei so that differential retention can be demonstrated. This differential retention leads us to conclude that the binding of H1 in chromatin changes with growth phase, being more easily dissociated with lowering pH in exponential than in stationary cell nuclei.

TABLE I

Relative Amounts (%) of Histone Fractions Retained in Nuclei Isolated from Exponential (E) and Stationary (S) L Cells with 1 m*M* Mg^{2+} and Ca^{2+}[a]

m*M* Citrate	3		5		10	
pH (calculated)	2.85		2.75		2.62	
Ionic strength ($\times 10^{-3}$)	6.9		7.3		8.6	
	E	S	E	S	E	S
Lysine-rich H1	21	20	8	16	6	8
Slightly lysine-rich H2a	15	16	17	16	19	17
H2b	27	29	33	32	33	31
Arginine-rich H4	18	19	20	19	22	21
H3	19	16	22	17	20	23

[a] All nuclei were isolated from monolayer cultures of mouse L cells in medium containing 0.1% Triton X-100 and 0.1 *M* sucrose, in addition to $MgCl_2$, $CaCl_2$ and citrate in quantities indicated above.

TABLE II

The Incorporation of [^{3}H]Lysine and ^{32}P into L Cell Histones in Either Exponential Phase (Exp) or Stationary Phase (G_0)[a]

Histone fraction	G_0 (cpm/μg)		Exp. (cpm/μg)	
	[^{3}H]Lysine	^{32}P	[^{3}H]Lysine	^{32}P
H1	80	9	259	29
H3	25	4	151	17
H2b	15	8	190	24
H2a	35	7	135	14
H4	23	4	40	8

[a] Histones were extracted from nuclei isolated in medium containing 0.1% Triton X-100, and 0.1 *M* sucrose 1 m*M* Mg^{2+} and 1 m*M* Ca^{2+} at neutral pH.

In order to investigate whether this preferential pH susceptibility could be ascribed to synthesis of new H1 molecules or to their degree of phosphorylation, we carried out pulse-labeling studies, using [^{3}H]lysine and ^{32}P. In this case, nuclei were isolated at neutral pH in order to ensure total conservation of H1. The specific activities for ^{3}H and ^{32}P of the five histone fractions recovered from either phase are illustrated in Table II (17). These results demonstrate the presence of newly synthesized histones in exponential cells, as evidenced by a threefold increase in specific activity of ^{3}H in lysine-rich fraction H1. This same fraction also shows an equivalent increase in phosphorylation. Presumably, newly synthesized highly phosphorylated H1 molecules are less tightly bound to DNA than the older less phosphorylated ones, and, therefore, the former can be lost during nuclear isolation under conditions where the latter are retained.

B. Arginine-Rich Histones: Extractability, Synthesis, and Acetylation

In contrast with lysine-rich histones, where pH appears to be the critical factor for retention, arginine-rich histones were found to be highly sensitive to changes in divalent cation concentrations. When nuclei are isolated in media containing chelating agents, such as EDTA or citrate in the presence of Mg^{2+} and Ca^{2+} in concentrations of 1 m*M* or less, considerable loss of the arginine-rich fractions occurs during isolation. Note the relative amounts of H3 and H4 retained under such conditions (Table I), as compared with those illustrated in Table III where the divalent cation concentration was raised from 1 to 10 m*M*, in the presence of variable amounts of citrate.

TABLE III

Relative Amounts (%)[a] of Histone Fractions Retained in Nuclei Isolated from Exponential (E) and Stationary (S) L Cells with 10 m*M* Mg^{2+} and Ca^{2+}[b]

m*M* Citrate	0		26		52		104	
pH (calculated)	7.00		2.37		2.22		2.09	
Ionic strength ($\times 10^{-3}$)	60.0		64.7		66.5		68.6	
	E	S	E	S	E	S	E	S
Lysine-rich H1	22 ± 5	22 ± 4	—	—	—	—	—	—
Slightly lysine-rich H2a	17 ± 2	16 ± 9	17 ± 1	16 ± 1	19 ± 4	20 ± 2	17 ± 1	20 ± 1
H2b	35 ± 2	32 ± 4	30 ± 1	30 ± 2	34 ± 1	31 ± 1	33 ± 1	33 ± 2
Arginine-rich H4	17 ± 4	11 ± 2	35 ± 1	26 ± 2	38 ± 2	22 ± 3	30 ± 3	22 ± 2
H3	10 ± 3	18 ± 4	19 ± 0	28 ± 3	20 ± 3	26 ± 4	20 ± 3	25 ± 3
Total arginine-rich H3/H4	27	29	54	54	48	48	50	48

[a] Standard deviations indicated.

[b] All nuclei were isolated from monolayer cultures of mouse L cells in medium containing 0.1% Triton X-100 and 0.1 *M* sucrose in addition to $MgCl_2$, $CaCl_2$, and citrate in quantities indicated above.

Maximum recovery of H3 + H4 was obtained when nuclei were isolated in the presence of 26 mM citrate, at a pH where all of the H1 is being lost (16). Here again, preferential retention of H4 over H3 or vice versa depends on whether cells are stationary (G_0) or in a logarithmic phase of growth. In all cases, exponential cell nuclei appear to retain more H4 than H3, while the opposite is true for G_0 cells. However, the total amount of H3 + H4 retained is the same for both cell phases, regardless of ionic conditions. We propose that this could be explained on the basis of the structural complementarity of the two arginine-rich histones in the formation of the nucleosome. It is conceivable that the orientation of the H3–H4 dimers can change as a function of the transcriptional activity of the nucleosomes such that H3 becomes more exposed and, consequently, more susceptible to dislodgement from transcriptionally active regions of the genome, while H4 becomes more exposed in inactive regions. The relative abundance of active versus inactive regions in cycling and noncycling cells could then determine the preferential retention of either of the two arginine-rich fractions.

We next tested the existence of a possible correlation between this preferential retention and the synthesis or acetylation of the two arginine-rich histones. In analogy with the effect of phosphorylation, it appeared probable that highly acetylated molecules might be more easily dislodged from chromatin, due to weaker polar interactions with DNA.

Pulse labeling of both mouse L cells and human WI-38 with either [^{3}H]-arginine or [^{3}H]acetate followed by isolation of nuclei in media observed to yield optimum amounts of arginine-rich fractions, resulted in the specific activities illustrated in Table IV (17, 18). In the case of the human cells, no change can be observed in the acetylation of H3, and only a slight increase in acetylation of H4 from G_0 to exponential populations. In mouse cells, H3 acetylation is even higher in G_0 cells. Therefore, acetylation levels do not appear to be the cause of the weaker binding of H3 in exponential cells. On the other hand, newly synthesized molecules in exponential cells could be less tightly bound in chromatin than older molecules. Note the higher ^{3}H-arginine specific activity of H3 in relation to H4 in exponential mouse and human cells.

With regard to acetylation, it might be argued that the specific activities of [^{3}H]acetate presented in Table IV may not be representative of the total if the conditions of nuclear isolation employed resulted in the leakage of the more highly acetylated species in spite of the fact that these conditions were found to result in highest quantitative yield of arginine-rich fractions. In order to test this possibility, we compared the acetylation levels of histones recovered from nuclei isolated in media containing variable amounts of divalent cations and citrate (17). Results show that the higher ionic strength low pH method not only results in better recovery of arginine-rich histones

TABLE IV

Acetylation and Synthesis of Arginine-Rich Histones in Mouse and Human Cells[a]

Histone fraction	G_0				Exp.			
	Mouse L cells (cpm/μg)		Human WI-38 (cpm/μg)		Mouse L cells (cpm/μg)		Human WI-38 (cpm/μg)	
	[^{3}H]Arg	[^{3}H]Acetate	[^{3}H]Arg	[^{3}H]Acetate	[^{3}H]Arg	[^{3}H]Acetate	[^{3}H]Arg	[^{3}H]Acetate
H3	50	20.8	55	11.3	238	13.8	190	11.5
H4	44	13	9	7.5	198	12.3	136	11.5

[a] All nuclei were isolated from monolayer cultures in medium containing 0.1% Triton X-100, 0.1 *M* sucrose, 10 m*M* $MgCl_2$, 10 m*M* $CaCl_2$, and 26 m*M* citrate, pH 2.4.

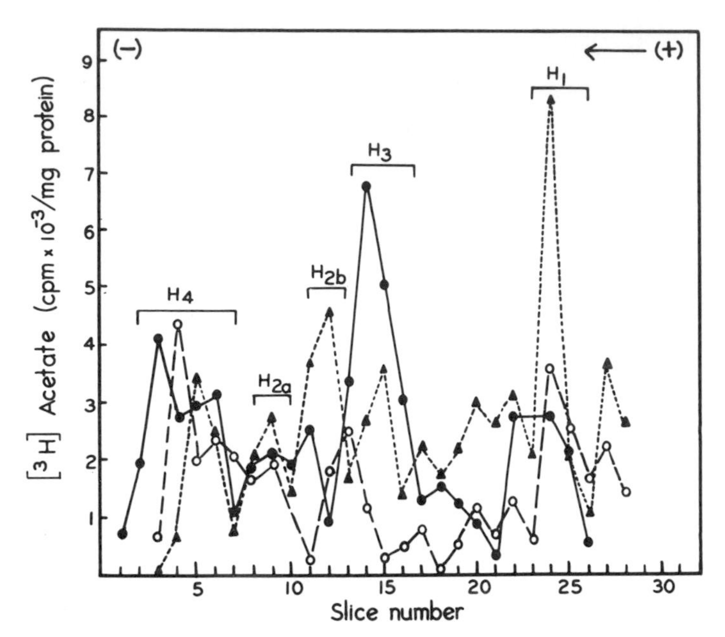

FIG. 1. The incorporation of [^{3}H]acetate into histones recovered from nuclei isolated in media at different pH and ionic strength and fractionated in acetic acid–urea gels according to the method of Panyim and Chalkley (19). ●, 10 m*M* Mg^{2+}, 10 m*M* Ca^{2+}, 26 m*M* citrate, pH 2.4. ▲ 10 m*M* Mg^{2+}, 10 m*M* Ca^{2+}, no citrate, pH 7.0. ○, 1 m*M* Mg^{2+}, 1 m*M* Ca^{2+}, 10 m*M* citrate, pH 2.6.

as seen in Table III, but also retains more of the acetylated H4 and H3, as illustrated in Fig. 1 (19). Therefore, one can rule out preferential loss of acetylated histones as the cause of the apparent lack of correlation between histone acetylation and differential retention in the nuclei of arrested and cycling cell populations.

III. CHANGES INDUCED BY SV40 TRANSFORMATION

These studies were carried out using monolayer cultures of both mouse 3T3 and human WI-38 fibroblasts before and after transformation with SV40 virus (18, 20, 21). Care was taken to ensure that both normal and transformed cells were in log phase at the time of labeling and harvesting. Cell-cycle analysis of log-phase normal and SV40-transformed WI-38 cells revealed no significant differences in mean generation time (20 hours) or length of S phase (11 ± 1 hour).

A. Nonionic Binding and HCl Extractability

In these studies, we investigated not only histone retention in nuclei isolated in different media but also the ease of extractability of each fraction from the isolated nuclei, using dilute mineral acid. By extracting nuclei with 0.25 *M* HCl for different time periods and quantitating the amounts of each fraction recovered, one can estimate the relative degree of polar versus nonpolar interactions of the histone proteins in chromatin. Table V illustrates the results obtained when using two different methods for nuclear isolation: one a low pH, high divalent cation method, and the other, a neutral pH, low divalent cation one aimed at conserving the H1 molecules (18). Extraction times of 2 and 20 hours with HCl were compared in each case. It can be seen that not only more arginine-rich histones are retained, using the high divalent cation method, but also that recovery increases with the longer extraction times. Note, however, that more H4 than H3 is recovered from SV40-transformed cells after 2-hour extraction, while more H3 than H4 is recovered from normal cells. After 20-hour extraction one does recover the two species in 1 : 1 proportion in both cases. Examination of the nuclear residues by SDS gel electrophoresis showed no evidence of any histones remaining; therefore, extraction times longer than 20 hours were deemed unnecessary. The above results indicate that H4 is more resistant to acid extraction in normal than in transformed cells, while the opposite is true for H3. This may be interpreted to mean that the higher resistance of

TABLE V

Relative Amounts of Protein in Histone Fractions Extracted from Nuclei of Normal and SV40-Transformed WI-38 Cells by the pH 2.4 and pH 7.2 Methods

Histone fraction	Nuclei isolated at pH 2.4 with 10 m*M* Mg^{2+} + Ca^{2+}, 25 m*M* citrate				Nuclei isolated at pH 7.2 with NaCl-EDTA			
	SV-WI-38 (%) Total histones		WI-38 (%) Total histones		SV-WI-38 (%) Total histones		WI-38 (%) Total histones	
	2 hr	20 hr	2 hr	20 hr	2 hr	20 hr	2 hr	20 hr
H1	—	3	—	4	18	16	15	16
H3	20	31	32	35	20	25	27	22
H2b	26	22	26	15	30	28	33	27
H2a	19	13	16	13	14	15	8	15
H4	35	31	26	33	18	16	17	20
H4/H3	1.75	1.0	0.81	0.94	0.90	0.64	0.63	0.91
Total H3 + H4	55	62	58	68	38	41	44	42

one or the other fraction to extraction with acid is due either to its being less accessible within the chromatin fiber or to extensive nonpolar interactions with other components of the chromatin or more likely a combination of the two. Here again, one could visualize a possible reorientation of the H3–H4 dimer within the nucleosomes of the transformed genome, which would make one of the two arginine-rich fractions more exposed in the chromatin of one cell type than in the other.

In contrast with arginine-rich histones, fraction H1 is presumed to have a much more exposed location in the chromatin fiber and to bind to DNA largely via salt linkages; therefore, it is not surprising that it comes out readily after short HCl treatment with no observable differences in recovery between normal and transformed cells.

B. Synthesis and Phosphorylation of Lysine-Rich Histones

Pulse-labeling experiments with ^{3}H-labeled amino acids and ^{32}P were carried out to investigate any alteration in synthesis or phosphorylation patterns of lysine-rich fractions occurring as a consequence of SV40 transformation in both mouse 3T3 and human WI-38 fibroblasts (18, 20, 21). For optimum recovery of the lysine-rich fractions, nuclei were isolated at neutral pH after pulse labeling. Results presented in Table VI, demonstrate an increased specific activity of H1 and H2a after SV40 transformation in both species. A parallel increase in phosphorylation of H1 is also apparent, although more dramatic in the case of the human cells. Phosphorylation of H2a, however, appears to decrease slightly in both cell types after transformation.

Since cell cycle parameters are very similar in both normal and transformed cells, and both cultures were harvested in log phase, it seems improbable that these changes are cell-cycle-dependent. The similarities of the results observed after SV40 transformation of two different mammalian species suggest, rather, an SV40-dependent stimulation of the phosphorylation of the lysine-rich H1 fraction.

C. Synthesis and Acetylation of Arginine-Rich Histones

For these studies, cells were pulse labeled with either ^{3}H-labeled arginine plus lysine or with [^{3}H]acetate, and nuclei were isolated using the low pH, high divalent cation method in order to ensure complete recovery of arginine-rich fractions (18). Table VII represents the results obtained from histones of normal and SV40-transformed human WI-38 cells. The rate of amino acid incorporation into histones other than H1, which leaks out of nuclei at this low pH, appears generally higher than in Table VI, attesting

TABLE VI

Synthesis and Phosphorylation of Histones in Mouse and Human Cells before and after Transformation with SV40 Virus[a]

Histone fractions	Mouse cpm/μg protein				Human cpm/μg protein			
	3T3		SV-3T3		WI-38		SV-WI-38	
	[^{3}H]Leu	^{32}P	[^{3}H]Leu	^{32}P	[^{3}H]Arg-Lys	^{32}P	[^{3}H]Arg-Lys	^{32}P
H1	243	272	310	328	167	55	z36h	
H3	626	—	476	—	50	—	104	—
H2b	584	—	422	—	29	—	112	—
H2a	330	120	750	80	72	51	270	47
H4	181	—	362	—	32	—	40	—

[a] Histones were extracted with 0.25 HCl for 20 hours from nuclei isolated at pH 7.2.

TABLE VII

Acetylation and Synthesis of Histones Extracted from Nuclei Isolated at pH 2.4

Histone fraction	SV-WI-38			WI-38		
	% Total histone	[^{3}H]Arg-Lys (cpm/μg)	[^{3}H]acetate (cpm/μg)	% Total histone	[^{3}H]Arg-Lys (cpm/μg)	[^{3}H]acetate (cpm/μg)
H3	31	445	23	35	373	11.5
H2b	22	306	—	15	350	—
H2a	13	228	—	13	270	—
H4	31	270	13.5	33	272	11.5

to the preservation of newly synthesized molecules in nuclei isolated through this method. Specific activities for histone fractions of normal and transformed cells are rather similar, with the exception of H3 which is higher in SV40-transformed cells. Acetylation is also different in the case of fraction H3, whose specific activity doubles after transformation.

Higher rates of synthesis and acetylation of arginine-rich H3 might possibly explain the difference in the HCl solubilization of H3 and H4 in normal and transformed cells. While newly synthesized and acetylated molecules tend to leak out of nuclei during cell lysis at low divalent cation concentration and neutral pH, once they are retained in nuclei isolated in optimal ionic strength conditions, these same histone species are the most resistant to HCl extraction. As a consequence, more highly acetylated H3 in transformed cells is also more resistant to HCl extraction than the less acetylated H3 of normal cells. We postulate that an optimum divalent cation environment during cell lysis is essential for the preservation of arginine-rich histone conformation in those portions of the molecules that are involved in hydrophobic interactions with other chromatin components. More highly acetylated molecules may have weaker polar interactions with DNA, and, therefore, their retention in chromatin depends largely on their nonpolar binding with each other and other components of the genome. The extent of hydrophobic interactions will in turn affect the extractability of these molecules with mineral acid.

IV. DISCUSSION AND CONCLUSIONS

The body of evidence presented above indicates that, although the role of histones in chromatin may be primarily a structural one, there are enough structural variations in the mode of association of histones with each other and with DNA or other components of chromatin to support the hypothesis that they also participate in genetic control.

Postsynthetic modifications of the histones, such as acetylation and phosphorylation, have long been suspected of allowing for such a role by modulating the interaction of histones with DNA (15). However, the lack of consistency in the results reported by various workers who tried to correlate modification reactions with genetic activation, particularly in the case of acetylation (17, 22–25), threw considerable doubt into such hypothesis. The finding of selective leakage of histones from nuclei during isolation, particularly in the case of the arginine-rich fractions at low ionic strength, raises the possibility that the reported inconsistency could result from the partial retention within isolated nuclei, of a variable population of arginine-rich fractions, depending on cell type and lysing conditions used in each study. Different states of genetic control are likely to be accompanied by conformational changes in the chromatin fibers involving increased exposure of one or more histone fractions. Since most aqueous methods currently used for isolation of nuclei do not preserve all of the histones (26), it is very difficult to interpret and compare many of the results. Once we become aware of this potential flaw, it is encouraging to ascertain that one can take advantage of this leaky system in order to correlate the extractability of histone fractions with the extent of their modification reactions and various states of genetic activation in chromatin.

Using such a system, we have been able to demonstrate that newly synthesized highly phosphorylated H1 molecules are less tightly bound to DNA than the older less phosphorylated species. It is well known that cells which are progressing through the cell cycle have higher levels of H1 phosphorylation than G_1-arrested populations (17, 27), and they transcribe 2–3 times as much RNA (28). The positive correlation between H1 phosphorylation and transcriptional activity can thus be explained on the basis of looser coils of nucleosomes in the formation of the 20-nm "native fiber" in interphase chromatin.

The orientation of the nucleosomes themselves could also change as a function of the transcriptional activity of the "native fiber" in chromatin. This is supported by our finding that H3 is more exposed than H4 and, consequently, more susceptible to dislodgement from transcriptionally active cycling cells, while H4 becomes more exposed than H3 in less active G_0 cells. Moreover, the accessibility of H3 and H4 to extraction by acid was also found to alternate in normal and SV40-transformed human cells. This alternating exposure appears to correlate with different levels of acetylation in H3 and H4 histones, supporting the concept of a structural change in the nucleosomes in regions of active transcription.

ACKNOWLEDGMENTS

My sincere thanks go to Dr. Gary Stein in whose laboratory I conducted some of the above work during my sabbatical year. I also acknowledge the National Research Council of Canada Grant A-4433 for financial support.

REFERENCES

1. Olins, A. L., and Olins, D. E., *Science* **183,** 330 (1974).
2. Kornberg, R. D., and Thomas, J. O., *Science* **184,** 865 (1974).
3. Kornberg, R. D., *Science* **184,** 868 (1974).
4. Thomas, J. O., and Kornberg, R. D., *Proc. Natl. Acad. Sci. U.S.A.* **72,** 2626 (1975).
5. Langmore, J. P., and Wooley, J. C., *Proc. Natl. Acad. Sci. U.S.A.* **72,** 2691 (1975).
6. Weintraub, H., Palter, K., and Van Lente, F., *Cell* **6,** 85 (1975).
7. Oudet, P., Gross-Bellard, M., and Chambon, P., *Cell* **4,** 281 (1975).
8. Finch, J. T., and Klug, A., *Proc. Natl. Acad. Sci. U.S.A.* **73,** 1897 (1976).
9. Varshavsky, A. J., Bakayer, V. V., and Georgiev, G. P., *Nucleic Acids Res.* **3,** 477 (1976).
10. Witlock, J. P., Jr., and Simpson, R. T., *Biochemistry* **15,** 3307 (1976).
11. Zimmerman, S. B., and Levin, C. J., *Biochem. Biophys. Res. Commun.* **62,** 357 (1975).
12. Kuo, M. T., Sahasrabuddhe, C. G., and Sanders, G. F., *Proc. Natl. Acad. Sci. U.S.A.* **73,** 1572 (1976).
13. Wigler, M. H., and Axel, R., *Nucleic Acids Res.* **3,** 1463 (1976).
14. Gottesfeld, J. M., Murphy, R. F., and Bonner, J., *Proc. Natl. Acad. Sci. U.S.A.* **72,** 4404 (1975).
15. Hnilica, L. S., *in* "The Structure and Biological Function of Histones," p. 79. CRC Press, Cleveland, Ohio, 1972.
16. Krause, M. O., Yoo, B. Y., and MacBeath, L., *Arch. Biochem. Biophys.* **164,** 172 (1974).
17. Krause, M. O., and Inasi, B., *Arch. Biochem. Biophys.* **164,** 179 (1974).
18. Krause, M. O., and Stein, G. S., *Exp. Cell Res.* **92,** 175 (1975).
19. Panyim, S., and Chalkley, R., *Biochemistry* **8,** 3972 (1969).
20. Krause, M. O., and Stein, G. S., *Exp. Cell Res.* **100,** 63 (1976).
21. Krause, M. O., Noonan, K. D., Kleinsmith, L. J., and Stein, G. S., *Cell Differ.* **5,** 83 (1976).
22. Allfrey, V. G., Pogo, A. O., Kleinsmith, L. J., and Mirsky, A. E., in "Histones" (A. V. S. de Rueck and J. Knight, eds.), p. 42. Churchill, London, 1966.
23. Allfrey, V. G. *in* "Histones and Nucleohistones" (D. M. Philips, ed.), p. 241. Plenum, New York, 1971.
24. Clever, U., and Ellgaard, E. G., *Science* **169,** 373 (1970).
25. Monjardino, J. P. P. V., and MacGillivray, A. J., *Exp. Cell Res.* **60,** 1 (1970).
26. Krause, M. O., *in* "Methods in Chromosomal Protein Research" (G. S. Stein, J. L. Stein, and L. J. Kleinsmith, eds.). Academic Press, New York (in press).
27. Hohmann, P. G., Tobey, R. A., and Gurley, L. R., *J. Cell Biol.* **251,** 3685 (1976).
28. Wong, K-Y., Patel, J., and Krause, M. O., *Exp. Cell Res.* **69,** 456 (1971).

5

Triggers, Trigger Waves, and Mitosis: A New Model

Patricia Harris

I. INTRODUCTION

A. The Concept of a Trigger

The occurrence in biologic processes of abrupt onsets or sudden changes in rate has prompted many biologists to use the term "trigger" in describing these events. In considering the nature of such triggered events as fertilization, muscle contraction, and the nerve impulse, T. H. Bullock (16) defined the essential common denominators in this class of phenomena as: "a) a pull, that is to say, an addition of energy to a system, at a rate which may vary within wide limits, b) a critical point, beyond which c) there is released a store of energy, d) in a manner which is independent of the time course and energy content of the pull and beyond the control of that input, i.e., is all-or-none."

The apparent sudden onset of mitosis and its tendency to go to completion once started, suggested to many cytologists that some trigger may be operating to set off the division process. Mazia (69), in his extensive review, has pointed out the difficulties in assigning a trigger to any one of the many overlapping preparatory steps occurring during interphase, and instead utilizes the concept of parallel pathways which eventually converge at prophase, a "point of no return." While the condensation of chromosomes in prophase is an important visible marker, certain changes in cell behavior in the transition period from interphase have caused several workers to propose an antephase or predivision period (cited by Mazia, 69, p. 151). Recent work now indicates that antephase may already be a stage of mitosis, and if a trigger is to be assigned to mitosis at all, it would be at the transition from G_2 to antephase. This idea and its implications for the process of mitosis, based on a large body of published observations and seasoned with speculations, will be developed in the following sections.

B. Current Models for Mitosis

In recent years the primary interest in mitosis has been in the mechanism of chromosome movement itself, and much has been learned about the disposition of the structural elements of the mitotic apparatus and their biochemical characteristics. Fine structure studies have identified microtubules, and possibly actin, as constituents of the traction system, and the nature and conditions for their polymerization and depolymerization have now been studied in some detail (121). Several models based on current knowledge of the traction system have been proposed for chromosome movement. Some assume the microtubules to be the force producers, either by polymerization and depolymerization (59), by sliding with respect to each other (76, 80) or lateral interaction or "zipping" (5). Others ascribe the force production to actin (98) or an actin–myosin complex (32) coupled to the spindle microtubules, whose depolymerization would regulate the speed of chromosome movement. None of these models, however, adequately accounts for the entire process, beginning with the initiation of growth of the mitotic apparatus and ending with its final dissolution.

Another element of the mitotic apparatus, and one which has certainly not received its due consideration in models for mitosis, is the mass of vesicles or elements of the smooth endoplasmic reticulum associated particularly with the polar regions in both animal and plant cells (for reviews, see 51, 55). As Forer (32) has pointed out, microtubules comprise only a very small fraction of the total mass of the *in vivo* mitotic apparatus. In cells with large asters, such as sea urchin eggs, the clear area of the mitotic apparatus devoid of yolk is mainly a mass of membrane-bounded vesicles (48, 49, 51). Isolated mitotic apparatus, prepared by a variety of methods (17, 33, 61, 100) consistently contain vesicles which are completely ignored or explained away as cytoplasmic contaminants. While some of the "contaminants" are obviously yolk or mitochondria, the great mass of vesicles remaining are without doubt remnants of the extensive system of smooth endoplasmic reticulum in the intact cell. Attempts to "clean up" preparations of isolated mitotic apparatus may be a classic example of "throwing out the baby with the bath water," so to speak. The membrane system may, in fact, be the long sought after polar anchors for the traction system and, more importantly, may provide the control mechanism for the mitotic process, as will be discussed later.

II. COMPONENTS AND CONTROLS

A. Microtubule Assembly *in Vitro*

Since microtubules were the first structures shown unequivocally to be a consistent part of the traction system, a search for mitotic control mechanisms has naturally focused on the requirement for microtubule polymerization. Biochemical studies on microtubules (for reviews, see 14, 81, 121, 124, 138) have often been confusing and contradictory, but from the accumulating mass of data, some idea of the properties of the "working" microtubule is beginning to take form. Some of the conditions for polymerization and, thus, possibilities for control are considered below.

1. Divalent Cations

Although purified microtubule subunit protein had been available for several years, Weisenberg (134) was the first to demonstrate repolymerization *in vitro* in solutions containing ATP or GTP, Mg^{2+}, and a good Ca^{2+} chelator. The low Ca^{2+} requirement suggested that this may be an important *in vivo* control of microtubule formation.

Other workers (47) confirmed Weisenberg's observations, showing inhibition of microtubule polymerization by Ca^{2+} concentrations considered within physiological range, 1–30 μM, and even lower when K^+ ions were increased to physiological concentration. The requirement for Mg^{2+}, the interaction between Ca^{2+} and Mg^{2+}, as well as total ion composition and pH, showed that the interrelationship of these factors strongly influenced the sensitivity of microtubules to calcium.

Olmsted and Borisy (81, 82), on the other hand, found in their preparations that calcium was inhibitory only at what they considered nonphysiological levels, on the order of millimolar concentrations, but was required at very low concentrations for microtubule polymerization. Mg^{2+} was found to be required in concentrations somewhat higher than Ca^{2+}, but, above 10 mM, it becomes inhibitory. Although the discrepancy between these results and those of Weisenberg is presumed due to Weisenberg's use of crude extracts, the differing results of Haga *et al.* (47) remain unexplained. Since the *in vivo* system involves more than simply purified brain tubulin, but also many other constituents that may affect the sensitivity of the system to calcium, one cannot rule out calcium as a possible regulator of microtubule polymerization.

2. Nucleotide Requirement

Based on the early study of the nucleotide chemistry of tubulin (136), the presence of GTP is known to be necessary for *in vitro* polymerization. Olmsted and Borisy (82) have shown that, in the polymerization process, GTP bound to tubulin is hydrolyzed to GDP. cAMP or cGMP or nonhydrolyzable analogues of ATP or GTP inhibit polymerization.

The possible role of cAMP as a mitotic control is suggested, especially because of its relationship to cytoplasmic Ca^{2+} concentrations (for reviews, see 27, 95). While its action may be a direct one on the polymerization of tubulin *in vitro*, cAMP also accelerates the efflux of Ca^{2+} from mitochondria (11). It also appears that cAMP stimulates microsomal Ca^{2+} pumps but inhibits sarcolemmal pumps (3, 25). Therefore, the effect of cAMP may be either to inhibit or stimulate microtubule assembly *in vivo,* depending on whether within the particular cell mitochondria or microsomes (smooth ER) are more important as Ca^{2+} regulators.

3. pH

The pH optimum for microtubule polymerization reported by various investigators ranges from pH 6.0 to 6.8. Haga *et al.* (47) showed that, as the pH is increased from 6.2 to 7.3, polymerization measured as increase in flow birefringence decreased to zero. The lower pH value correlates well with studies on the stability of isolated mitotic apparatus (61). It, thus, appears that cytoplasmic pH might also act as a regulator of mitosis.

4. Tubulin Concentration

Borisy *et al.* (10) have shown that a critical concentration of approximately 0.2 mg/ml was required for polymerization *in vitro*. Release of active tubulin from an inactive storage form, thus, could be a means of microtubule control *in vivo*.

5. Nucleating Centers

Observations by Borisy and Olmsted (9) and later studies by Kirschner *et al.* (64) showed that microtubule assembly required components larger than the 6 S dimers. Electron microscope studies showed this 30–36 S fraction to be composed of rings or discs, which are probably intermediates of the polymerization process.

At the cellular level, organizing centers appear to be required for most of the microtubule polymerization. Isolated aster centers from clam eggs (135), kinetochores on isolated chromosomes (129), flagellar fragments (109), centrioles (120) are all apparently capable of initiating assembly of brain tubulin preparations, even those lacking the 36 S "seed" and, thus, incapable of spontaneous nucleation.

6. Structural Polarity and Directional Growth of Microtubules

In studying using DEAE cellulose-decorated microtubule fragments, as well as flagellar fragments, Allen and Borisy (2) showed that at low concentrations of tubulin, growth occurred only at one end, and, only at high concentrations, could polymerization at the other end be induced. Similar observations on flagellar fragments by Rosenbaum *et al.* (109) also showed biased growth in favor of the distal end. Bloom and Bryan (8), using computer analysis of polymerization rates and final polymer concentrations, showed that experimental data fit a model of end-on addition of dimers, rather than insertion along the length of a microtubule.

7. Miscellaneous

Many other observations on the biochemistry of tubulin have been documented (121). Phospholipids, protein kinase, ATPase, heavy molecular weight components, and phosphorylation of tubulin have all been reported, but as many of the authors have pointed out, the results of these studies are greatly affected by the history of the tubulin being used; for example, its source, methods of purification, buffer used, length of storage, and innumerable unknown factors. The great amount of "inactive" tubulin that is thrown away because it refuses to behave as it "should" during the purification cycles of cold depolymerization alone gives cause for caution. Whether any of the above observations will have any meaning in terms of possible control mechanisms is too early to say. It is especially premature with regard to calcium, since there is increasing evidence for membrane-associated calcium sequestering systems related to motile activities in many nonmuscle cells.

B. Control Mechanisms *in Vivo*

Mitotic inhibitors have long been used as probes to study the requirements for and the reactions of cell division in living cells. Some of these, like dinitrophenol, are respiratory blocking agents, reducing the energy supply, while others, like colchicine, directly attack the spindle structure itself. On the other hand, substances such as D_2O (59) and various glycols (101) have been shown to augment existing spindles, both in birefringence and numbers of microtubules. Recently, Rebhun *et al.* (102), using a combination of augmentation by glycols, and metabolic inhibitors that normally slowly bring about the reduction in size of the mitotic apparatus (dinitrophenol, caffeine, and other methylxanthines), were able to balance the two effects. Their explanation for this steady state is that there is a continuing cycling of subunits through the mitotic apparatus.

Further studies by Rebhun's group (78, 79) suggest that the effect of methylxanthines is mediated through inhibition of glutathione reductase activity, since reduced glutathione could reverse the inhibitory effect. On the other hand, Weber (131) has shown that the effect of caffeine on the sarcoplasmic reticulum is an uncoupling of the Ca^{2+} pump from ATP hydrolysis, allowing release of Ca^{2+} from the sarcoplasmic reticulum. Therefore, the methylxanthines may also be affecting a calcium-sequestering system.

Mercaptoethanol, also a sulfhydryl-reducing agent, has itself been shown to be a mitotic blocking agent (72, 73). Sea urchin eggs, blocked with mercaptoethanol until the controls reached the four-cell stage, divided directly from one to four cells on recovery in sea water. Electron microscope studies of these blocked cells (52) showed that in cells blocked just prior to metaphase, the spindle shortened to less than half its original length, while the cell centers at each pole separated to form four equidistant poles, each connected by a new spindle. It, thus, appeared that microtubule polymerization and depolymerization were occurring simultaneously in the aster centers and were probably two entirely different reactions, normally separated either in space or in time. It should be noted also that once the metaphase chromosomes had separated, mercaptoethanol no longer blocked mitosis, but, it fact, seemed to speed up the anaphase movement.

The ability of the cells to form a quadripolar figure in the block was greatly diminished in cells blocked at earlier stages, suggesting that there is an increasing availability of tubulin from an inactive pool as division progresses. Such a conclusion is consistent with studies by Stephens (125) on the sea urchin *Strongylocentrotus droebachiensis,* showing that the mobilization of previously unavailable monomer to an active pool occurs at one discrete time in the cell cycle, with the amount being determined by the temperature during early prophase. A study by Sluder (119), using Colcemid blocking followed by UV release of the block to manipulate experimentally the available tubulin pool in eggs of the sea urchin *Lytechinus variegatus,* showed that spindle size and growth rate are determined by factors other than the maximum rate of tubulin polymerization in the cell. Three factors were suggested as possible controls: modulation of the pool of polymerizable tubulin, regulation of intracellular ionic or nucleotide concentrations, or modulation of the nucleating ability of the kinetochores and centrosomes.

In a structure as transitory as the mitotic apparatus, it is probable that not just one, but several of the many previously mentioned possible control factors are operating. At the present time, however, there is considerable evidence in many types of cells that there is a mechanism for controlling calcium ions, which could have a direct effect on microtubule polymerization or which could act indirectly through colligative action with other control factors. Such a control mechanism would also be important in the regulation of contractile proteins, since the interaction of actin and myosin, as well as the state of actin polymerization itself, is a calcium-dependent function.

C. Contractile Proteins

The presence of actin, myosin, and a range of other muscle proteins in nonmuscle cells is now a well-established fact (reviewed by Pollard and Weihing, 90). What is less well established, and currently hotly debated, is the presence of these proteins in the mitotic apparatus and whether they play a role in chromosome movement. There is evidence for the presence of actin filaments in the spindle, based on the technique developed by Ishikawa *et al.* (60) for heavy meromyosin labeling of actin (for reviews, see 32, 57) but there are certain drawbacks to this technique. The cells must be broken or glycerinated to permit the entry of the heavy meromyosin, and this process itself may cause considerable translocation or loss of cell contents. Furthermore, Yagi *et al.* (141) have shown that heavy meromyosin may catalyze the polymerization of G actin to F actin.

Using fluorescent-labeled heavy meromyosin for light microscopy, Sanger (113, 114) has reported the presence of actin in the kinetochore fibers of the spindle. This technique has some advantage in that the distribution of labeled material can be observed in the entire cell, but since these cells, too, must be glycerinated, the same objections apply as before. Indirect immunofluorescent localization of actin (65), which also allows light microscopic examination of whole cells, has not yet demonstrated actin in the spindle, although McIntosh *et al.* (75, p. 37) report some evidence for actin at the poles of ptK$_1$ cell spindles. There is some question of the specificity of antibodies to such highly conserved and, thus, weakly antigenic proteins as actin, and results with these methods should still be interpreted with caution.

A present controversy has arisen over the nature of the components giving rise to the birefringence of the mitotic apparatus, and the future of several models for chromosome movement depend on its outcome. Following the lead of Inoué and Sato (59), a number of workers (111, 112, 115, 126) argue that microtubules alone are responsible for the birefringence. On the other hand, Goldman and Rebhun (45) showed that 30–50% of the birefringence of the isolated spindles remained, even when microtubules had broken down and only aligned particles were present. Forer and Zimmerman (33–35) developed an improved method of isolating the sea urchin mitotic apparatus, and Forer *et al.* (36) have now demonstrated that 45% of the isolated spindle birefringence is due to nontubulin and nonmicrotubule components. This material, or γ-component, was found to be extractable with 0.5 *M* KCl. The remaining components after KCl extraction were identified on SDS gels as an actin band that represented roughly 2% of the total protein, tubulin, and a high molecular weight component. The nature of the γ-component, and what role, if any, it plays in chromosome movement has yet to be elucidated.

It is interesting to note that the response of actin and myosin to calcium concentrations seems to be the reverse of that of tubulin. High concentrations of calcium bring about the depolymerization of microtubules, while actin and myosin exhibit a heirarchy of responses to decreasing calcium concentrations. In a series of studies on amoeboid movement, Taylor *et al.* (128) and Condeelis *et al.* (18) found that associated actin and myosin contracted above a threshold concentration of about 7.0×10^{-7} *M*. At subthreshold concentrations in the presence of Mg^{2+}–ATP, actin and myosin were dissociated, and streaming was observed. With increasing amounts of EGTA, which lowered the calcium concentration still further, the actin filaments, and, to a lesser extent, the myosin filaments were depolymerized. These observations should be kept in mind when trying to preserve these proteins in isolated mitotic apparatus or in fixed preparations for electron microscopy. While it is very likely that actin, myosin, and tubulin are all present in the mitotic apparatus in the living cell, it is very unlikely that they will be found in their polymerized form at the same time or in the same place in such preparations.

D. Membranes and Ca^{2+} Regulation in Motile Systems

In muscle, where calcium plays an important role in the contractile process, the sarcoplasmic reticulum possesses a calcium pump driven by a Ca^{2+}-activated ATPase and functions to control the cytoplasmic calcium ion concentration by sequestering or releasing Ca^{2+} from its cisternae (reviewed by Hasselbach, 54). Structures analogous to the sarcoplasmic reticulum have been demonstrated in other contractile or locomotory cells, where vesicles have been shown to sequester calcium. For example, the contractile ciliate *Spirostomum* has a filamentous contractile system associated with vesicles shown to accumulate calcium oxalate precipitates (31). The amoebae *Amoeba proteus* and *Chaos chaos* also have an ATPase-sensitive membrane system capable of accumulating calcium, as demonstrated by calcium oxalate precipitates (103).

In a study of human granulocyte chemotaxis, Gallin and Rosenthal (43) provide evidence for an association between calcium, apparently regulated by a calcium-sequestering system, and microtubule assembly. Ridgway and Durham (104), using the Ca^{2+}-specific photoprotein aequorin injected into the slime mold *Physarum polycephalum,* showed oscillations of calcium ion concentrations related to shuttle streaming. In another study of shuttle streaming in *Physarum,* Braatz (12) claims to have identified a vacuolar calcium pumping system by histochemical localization with oxalate. In plasmodial regions of protoplasmic influx, calcium precipitates are predominantly localized within vacuoles, while, in regions of protoplasmic outflow, they are mainly scattered free in the gound cytoplasm. In *Chironomus* salivary gland cells injected with aequorin, Rose and Lowenstein (108) have demonstrated the presence of an energized calcium-sequestering system. A membrane fraction prepared by differential centrifugation of platelet homogenates has a high ATP-dependent, oxalate-enhanced calcium uptake activity similar to muscle sarcoplasmic reticulum (107).

Recent use of Ca ionophores, which equilibrate cytoplasmic calcium ion concentration with that of the medium, further suggests a role for calcium ions in the control of microtubule polymerization. Schliwa (116), using the heliozoan *Actinosphaerium eichhorni* and the divalent cation ionophore A23187, showed that in the presence of the ionophore, microtubule polymerization can be controlled by varying the amounts of external calcium ion. He concluded that microtubule formation *in vivo* is probably regulated by micromolar amounts of calcium ion. In mammalian tissue culture cells, Fuller *et al.* (40), using indirect immunofluorescent labeling of tubulin to follow the effect of ionophore A23187, demonstrated a rapid depolymerization of cytoplasmic microtubules.

Criticism of the results with ionophores because of their possible side effects was circumvented by Kiehart and Inoué (63), who injected $CaCl_2$ directly into dividing echinoderm eggs. The calcium produced only local depolymerization, even when large amounts were injected, which would be expected to depolymerize microtubules *in vitro* if there were free diffusion. This suggests that there is a rapid sequestering of calcium ions at the time the spindle is formed and that calcium may locally control microtubule assembly and disassembly.

Further evidence for a calcium-sequestering system during mitosis was the identification of a Ca^{2+}-activated ATPase in the isolated mitotic apparatus by Mazia *et al.* (74) and the demonstration that this enzyme activity is related to the cell cycle (85–87, 89). Thus, the presence of a massive system of vesicles, which makes up the greater part of the mitotic apparatus in sea urchin eggs and which is closely related to the growth of the microtubules, suggests that it may also play a role similar to the sarcoplasmic reticulum in regulating calcium ion concentration in the cytoplasm.

III. MEMBRANES IN THE MITOTIC APPARATUS

A. General Occurrence

Although the model for mitosis to be considered later is based primarily on observations of the first cleavage division in sea urchin eggs, the presence of membranes and vesicles at spindle poles is by no means limited to sea urchins. It is a general phenomenon in a wide range of animals and plants as well. While there is usually a greater accumulation of vesicular material in large cells, such as marine invertebrate eggs, where the asters are exaggerated, even in smaller cells that show an anastral type of spindle, there are always vesicles in the polar regions (reviewed by Harris, 51).

Porter and Machado (92) were the first to demonstrate the presence of the endoplasmic reticulum in the spindle and at the polar regions of dividing onion root tip cells. Since that time, the presence of vesicular elements at the poles of plant mitotic and meiotic spindles has been described in an ever increasing number of species. This work has been reviewed in detail by Hepler (55).

B. Sea Urchin Mitosis

As early as 1961, fine structure studies of mitosis in sea urchin eggs (48, 49) showed that the clear area of the mitotic apparatus seen in living cells, the so-called mitotic gel, was, in fact, made up largely of membrane-bounded vesicles, within which the microtubular system was embedded. In recent years, as information concerning the chemistry of microtubule protein and the conditions for its polymerization has become available, the possible role of this vesicular system as a regulator of calcium ions and as an anchor for the mitotic traction system has taken on greater significance. In a more recent study, using both light and electron microscopy, a close relationship between the vesicular system and microtubules was found to exist throughout the mitotic cycle (51).

Sectioned and stained material examined by light microscopy showed a steady increase in the size of the asters from early prophase through telophase. This sequence is shown diagrammatically in Fig. 1. The outer limits of the aster grow continuously from prophase through division. They reach the polar surfaces of the egg at anaphase, and, by late telophase, the astral rays have invaded the entire egg. The centrosphere region stains densely for lipid. In detergent-isolated mitotic apparatus, shown in Fig. 2, the centrosphere appears as a less dense region into which the spindle and astral fibers appear to insert and end. Within this fibrous less dense region is an accumulation of dense amorphous material, which is usually eccentrically displaced towards the ends of the spindle. The centrosphere regions with their clumps of dense material also grow continuously in size, showing no reversal, even though the chromosomes during this period have been moved first to the metaphase plate and then to their respective poles.

Fine structure studies show the aster to contain two main structural components: microtubules and an extensive system of smooth endoplasmic reticulum (Fig. 3), the latter particularly concentrated in the aster center. The astral rays, which radiate from the centrosphere toward the cell periphery, are composed of elongated vesicles accompanied by bundles of microtubules. At the point where the astral rays and spindle microtubules insert

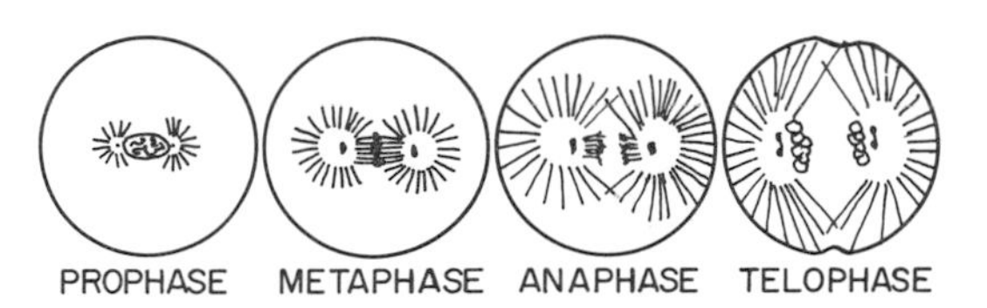

FIG. 1. Stages of mitosis in the sea urchin egg, showing continuous growth of both the aster and the centrosphere throughout the division process.

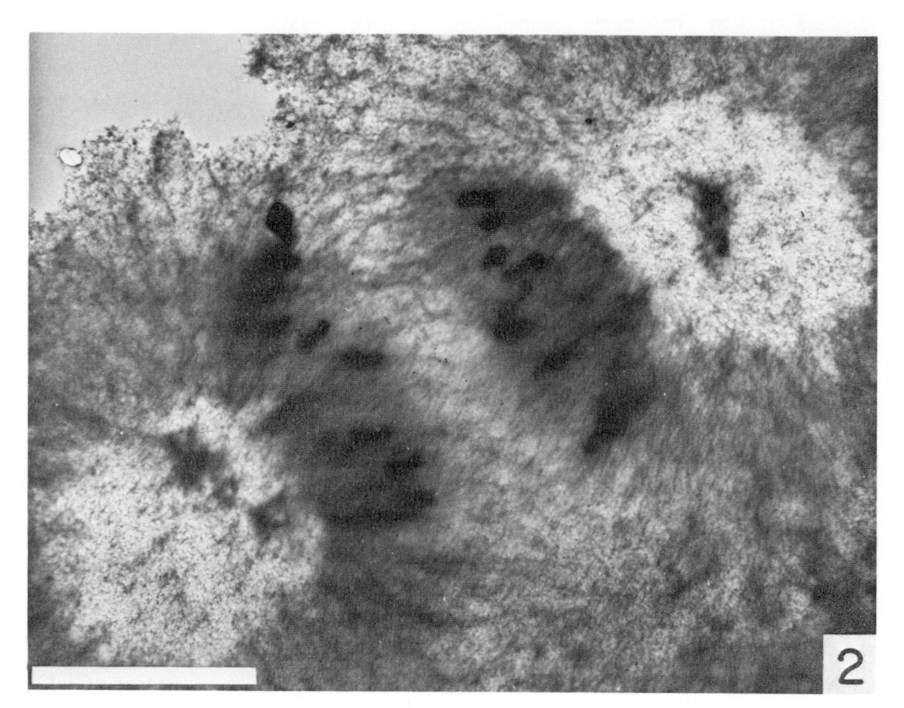

FIG. 2. Low-power electron micrograph of a detergent-isolated sea urchin mitotic apparatus. Spindle and aster fibers appear to end at the periphery of the centrosphere (51). Marker indicates 10 μm.

into the periphery of the centrosphere, the microtubules do not end, but rather show an abrupt configurational change (Figs. 4 and 5). The region into which the more orderly arrays of microtubules apparently disappear reveals a network of randomly oriented short segments of microtubules, or parts of longer lengths delimited by the section thickness. Often these segments are bent or distorted. The clumps of dense material associated with the centrioles show no definite structure at magnifications up to 80,000×, but usually very short segments of microtubules are embedded in or closely associated with them (Fig. 6).

IV. THE "CENTRAL APPARATUS" AND MITOTIC ASTERS

A. Homology of the Two Structures

The association of centrioles, Golgi vesicles, and what appeared to be radiating fibrillae or lengths of microtubules in interphase cells, has been referred to as the central body, central apparatus, microcentrum (137), the cell center, cytocentrum, or centrosphere (91), and, more recently, as the cytoplasmic microtubule complex (CMTC) (40). In view of the unexpected size and complexity of this organelle, and with some deference to its historical background, I am using the term "central apparatus" as more appropriate for the entire structure, and certainly more descriptive than the acronyms so much in vogue these days, and "cell center" or "centrosphere" for the innermost part of that structure.

Little is known about the role of the central apparatus in interphase cells. Wilson (138) notes that the center often forms a focus about which are aggregated certain of the other formed elements of the cytoplasm, such as Golgi bodies and mitochondria. Other granules, such as lysosomes, also may be found aggregated at the periphery of the center. Fine structure studies (91) show the center itself to be a highly differentiated zone of relatively high viscosity, from which larger cell organelles such as mitochondria and rough-surfaced endoplasmic reticulum are excluded. In animal cells, it

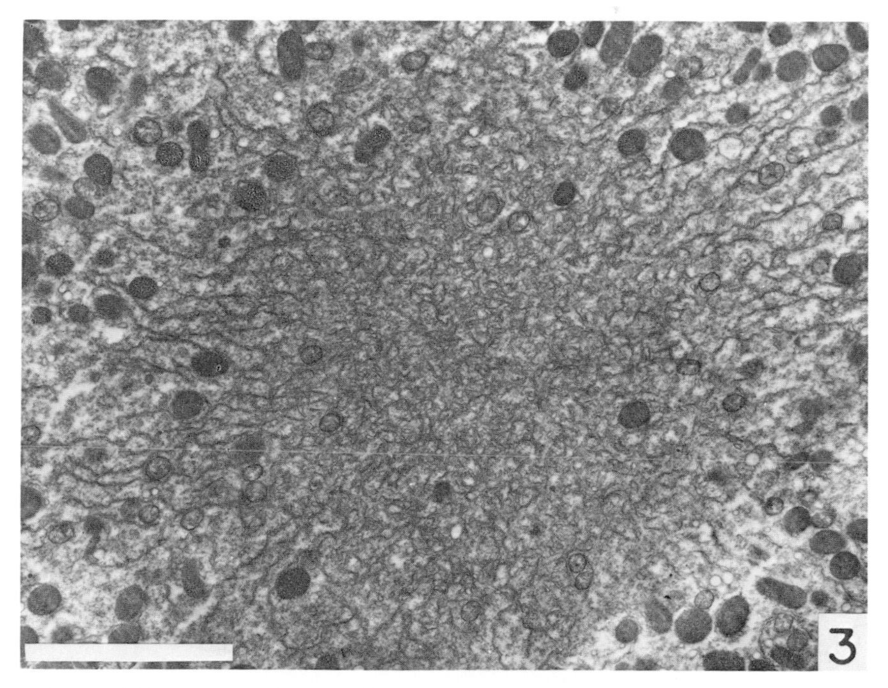

FIG. 3. Thick section through the edge of a growing aster, showing the mass of membranes and membranous extensions that form the astral rays. At this low magnification, the microtubules are not visible (49). Marker indicates 5 μm.

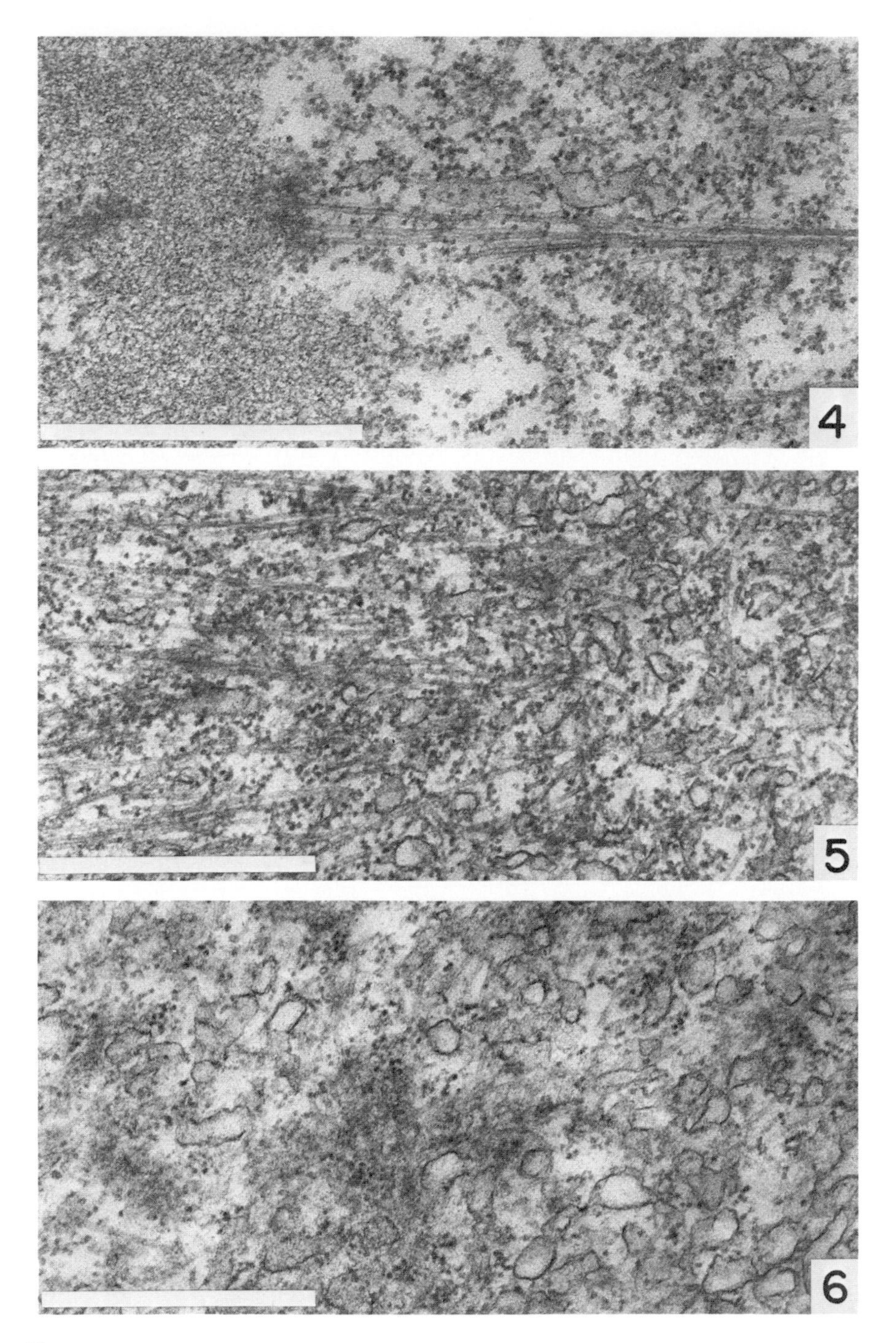
4
5
6

contains a pair of centrioles and dense amorphous material, which may be organized as centriolar satellites. This inner region is surrounded by membranous elements of the Golgi complex. Microtubules radiate out into the cytoplasm from the vicinity of the centrioles, with their proximal terminus usually in the dense pericentriolar material.

Although particle movement and streaming had been described previously in this central apparatus (26, 97), the first demonstration that these movements were associated with an extensive array of microtubules was that of Freed and Lebowitz (38). Using living HeLa tissue culture cells and sectioned material for electron microscopy, they showed that the paths of granules undergoing saltatory movements coincided with the arrangement of cytoplasmic microtubules. When the cells were treated with colchicine, the movement ceased and aggregated granules dispersed. These authors concluded that this structure may be a persisting form of the aster found in mitotic cells and may constitute an intracellular transport system of wide occurrence. Among such examples are the movement of pigment granules in melanophores and transport of secretory products in various cells.

With the development of techniques for obtaining reasonably specific rabbit antibodies to purified tubulin (21) and subsequent use of fluorescein-labeled goat antibody against rabbit γ-globulin (132), it was possible to visualize tubulin containing structures in whole fixed cells. Earlier attempts using immunologic techniques succeeded only in labeling the mitotic structures and vinblastine-induced paracrystals (21, 23, 39), but, with improved fixation methods (132, 133), the large mass of interphase cytoplasmic microtubules became visible. Similar studies by Fuller *et al.* (41) and Brinkley *et al.* (13) also showed that this structure broke down when cells entered mitosis, and all the fluorescence appeared in the spindle and associated asters. At late telophase or early G_1, cytoplasmic tubules reappeared, growing out from the centrosphere region toward the cell periphery. These authors suggested that there was a cyclic transition of tubulin from one structure to the other.

The similarity of particle movement found in this "central apparatus" of interphase cells to that of the mitotic spindle and asters is very striking. It has even been suggested that chromosome movement is a special case of this general microtubule-associated movement (4). Rebhun (for review, see

---

FIGS. 4–6. The microtubular traction system from chromosome to pole. Markers indicate 1 μm.

FIG. 4. Chromosome with kinetochore microtubules.

FIG. 5. Transition from outer part of aster and spindle (left) to centrosphere (right). Note the abrupt disorientation of the aligned microtubules as they enter the centrosphere.

FIG. 6. Amorphous dense material within the centrosphere, with numerous small pieces of microtubules.

99) has described the movement of particles stained with vital dyes, which move into the astral regions of many marine invertebrate eggs during mitosis. While there is a counter movement of pigment vesicles to the cell cortex in fertilized *Arbacia* eggs, this movement is insensitive to colchicine, but is stopped by cytochalasin B. In the sea urchin *Strongylocentrotus purpuratus,* pigment migration to the cortex is also seen but is apparently independent of the asters, since mitochondria are moved to the aster centers and accumulate at the periphery of the centrosphere (53). In Chinese hamster ovary cells, virus particles, which are known to associate with microtubules (20, 22), are found aggregated at the centers (46). Many other examples are cited by Bajer and Molé-Bajer (7) in their extensive review.

In plant cells, such poleward movement of particles has long been known from the work of Bajer and Molé-Bajer (6, 7) on *Haemanthus* endosperm. Acentric bodies, consisting of various granules and chromosome fragments, begin moving to the poles as early as prometaphase and metaphase. In methanol-treated endosperm, Molé-Bajer and Östergren (cited by Östergren *et al.,* 84) found the formation of multipolar spindles. In one cell, where one of these asterlike poles was clearly visible, small granules accumulated by moving to the pole from all directions. Careful measurements were made with frame-by-frame analysis of time-lapse films.

Electron microscope studies of *Haemanthus* endosperm cells blocked in mitosis with isopropyl *N*-phenylcarbamate (IPC) (56) also show multipolar spindles, but the micrographs give little clue as to what the organizing center might be. Numerous vesicles are present at the center of the radiating microtubules, and it may be that the Golgi apparatus or vesicular system is involved here, as it seems to be in animal cells.

B. Mitosis: Two Asters from One

It is obvious that there are many similarities between the single interphase aster and the pairs of mitotic asters, both in structure and function. The single interphase structure selectively moves particles to its center. When two asters are present and coordinated with nuclear events, chromosomes as well as other particles are moved to the aster centers. However, an interphase cell entering mitosis with an aster of the size and complexity we now know exists, would be faced with a serious problem of transforming its single aster into a bipolar mitotic apparatus. Obviously, a simple solution is to dismantle the old one and start over with two nucleating centers instead of one.

The work of Fuller *et al.* (40) shows that this is probably accomplished by a release of intracellular calcium. Using indirect immunofluorescent labeling of microtubules, they showed that the cytoplasmic microtubule complex or central apparatus is normally first assembled in early G_1 of the cell cycle and persists until G_2 or early prophase, at which time it rapidly disappears. Shortly afterward, the spindle apparatus begins to form. Using the ionophore A23187, they demonstrated that calcium influx could break down this extensive aster, suggesting that the normal first step in mitosis is a release of calcium. The obvious function would be to break down structures that might physically interfere with mitosis and that would at the same time make tubulin dimers available for the formation of the mitotic apparatus.

It has been proposed that during formation of the mitotic apparatus, intracellular calcium regulation is brought about by sequestering of calcium within a membrane system closely associated with the growing asters, and that breakdown of the asters is accompanied by calcium release (51). If an interphase aster is also formed and broken down, one would expect to find two periods of calcium release, each followed by sequestering in the cell cycle. Petzelt (85, 86) has found that a Ca^{2+}-activated ATPase in sea urchin eggs has two peaks in the cell cycle, the first occurring during interphase and the second during metaphase, which would coincide roughly with the proposed periods of low cytoplasmic calcium. It suggests that the ATPase involves a membrane-associated calcium pump. Some supporting evidence comes from earlier work of Heilbrunn on viscosity changes in the cytoplasm during the cell cycle (reviewed by Mazia, 69). Viscosity measurements were made on various marine eggs by centrifugal sedimentation of cytoplasmic particles. Typical changes were found in those of *Chaetopterus*, which showed low viscosity during interphase and a sharp increase just prior to spindle formation (the so-called mitotic gelation). Viscosity dropped rapidly during spindle formation to another low at metaphase, after which it again increased. Heilbrunn attributed viscosity changes to shifts in calcium ion concentrations, and indeed, the viscosity curve is almost the inverse of the Ca^{2+}–ATPase activity.

The surprising drop in cytoplasmic viscosity at the time a complex structural element is being formed within the cell caused Hiramoto (58) to investigate regional differences in cytoplasmic viscosity of fertilized sea urchin eggs during the period from fertilization through first division. By measuring the response of an iron particle in the cytoplasm to the application of magnetic force, he found some striking changes. Immediately after fertilization, there is a rapid increase in the viscosity, reaching a peak within 2 minutes. This is followed almost immediately by a rapid drop in viscosity, and, at 4 minutes, it reaches a minimum value lower than that found in the unfertilized egg. Following this low viscosity is a gradual increase which presumably is correlated with the formation of the sperm monaster and later the mitotic apparatus. Movement of the iron particle across the aster region of a mitotic cell showed a very high viscosity within the aster center, which dropped to a minimum in the region between the aster and cortex, and again reached a very high level in the cortex itself. The very lowest viscosity was shown to exist in the region between the separating sets of anaphase chromosomes.

Since the viscosity changes could not all be correlated with visible changes in structural organization, such as the formation of the monaster or the mitotic apparatus, Hiramoto concluded that changes in the invisible cytoplasmic network contributed to the mechanical properties. Our present knowledge of the ubiquitous occurrence of actin and myosin in nonmuscle cells and their role in contraction and motility (reviewed by Pollard and Weihing, 90) supports such a conclusion.

C. Fertilization: One Aster from None

The interaction of actin and myosin in the presence of calcium, therefore, could account for the increased viscosity following fertilization, since it has been shown that calcium is released from an internal store at the time of fertilization (68) and a calcium ionophore A23187 can bring about artificial activation of sea urchin eggs (122). More recently, Ridgway *et al.* (105) reported what they described as an explosive release of calcium in eggs of the killifish *Medaka* at fertilization. Unfertilized eggs were injected with the photoprotein aequorin, which luminesces in the presence of free calcium ions. At activation, the luminescence increased to 10,000 times the resting level for several minutes and then declined, with a half-life of about 100 seconds until it again reached the unfertilized level. The pattern of luminescence, traveling as a wave on the cell surface from the point of sperm entry, appeared as a rather narrow luminescent band, indicating that the released calcium was being rapidly removed from the cytoplasm. If calcium removal following its initial release is a general phenomenon of egg activation, it could account for the viscosity drop observed by Hiramoto (58). It might also account for the inability of activated sea urchin eggs to develop normally in the presence of ionophore A23187 (122), since the ability of the cells to regulate cytoplasmic calcium would be lost or inhibited.

The similarity between the reactions occurring at fertilization and those occurring at mitosis, i.e., a rapid release of calcium followed by its subsequent removal from the cytoplasm, may not be a mere coincidence. The fertilization reaction precedes the formation of the sperm aster, with its large numbers of microtubules (67). Later, the streak stage replaces the monaster prior to the beginning of prophase. On the time scale of mitosis, the streak stage would very likely represent Heilbrunn's preprophase mitotic gelation, but the existing measurements of viscosity and observations of structural changes are not timed precisely enough to draw any conclusions in this regard. The low cytoplasmic viscosity at metaphase probably reflects the active removal of calcium known to take place at that time (63), and the subsequent dissociation of actin and myosin.

D. Some Cycles within Cycles

A careful study of *Arbacia* mitosis by Longo (66) leaves no doubt that there are profound changes in cytoplasmic membranes at the time the asters begin to form and to ultimately incorporate the structure of the streak. Annulate lamellae, usually associated with "heavy bodies" in unfertilized as well as early fertilized eggs (1, 50, 130), and found in great numbers in the clear streak region, gradually lose their annulae and become indistinguishable from the smooth endoplasmic reticulum (ER) in the growing asters. At this time, the nuclear membrane also breaks down in very much the same way. The fact that many of these annulate lamellae reappear after mitosis is complete and daughter nuclei have reformed, suggests that cytoplasmic conditions have returned to the state existing prior to prophase.

Recently, it has been suggested that the formation of annulate lamellae is induced by the disintegration of microtubules (24). These workers found a sequence of long-term responses to antimitotic agents, which brought about complete disappearance of microtubules: first, a dispersion of the Golgi organelles, followed by an accumulation of smooth ER around the Golgi regions, a gradual hypertrophy of rough ER and finally the appearance of stacks of annulated lamellae. The demonstration of tubulin associated with the pores of annulate lamellae by Dales *et al.* (23), using ferritin-labeled antibody, and the selective binding of radioactive colchicine to nuclear membrane (37) suggest that the annulate lamellae may be a storage form of tubulin.

Annulate lamellae are commonly found in mature egg cells or rapidly proliferating cells of various types (for reviews, see 62, 140). In sea urchin eggs, great numbers of annulate lamellae are found associated with aggregates of small granules, which cytochemical studies indicate contain RNA and basic proteins (19). It has been suggested that these structures may be the site of the so-called "masked messenger" RNA stored in the egg during oogenesis (50). Since studies by Raff *et al.* (93, 94) have identified tubulin as one of the first proteins synthesized by the maternal messenger RNA, heavy bodies should be considered as a possible source of tubulin released by the fertilization reaction.

In a series of very interesting studies on partial activation of unfertilized sea urchin eggs, it has been shown that many of the "late" (after 4–5 minutes) changes following fertilization can be induced by NH_4OH. Among these are K^+ conductance (123), DNA synthesis (71), chromosome condensation (70), mRNA polyadenylation (139), protein synthesis (30), and initiation of the Ca^{2+}–ATPase cycle (88). Significantly lacking, however, is the respiratory burst, acid production, and calcium release of the "early" response, and the subsequent lack of a mitotic apparatus at the time of chromosome condensation and nuclear membrane breakdown. Obviously, stored tubulin known to exist in the unfertilized egg is not released, and evidence points to calcium as the most likely factor in this process (30).

The lack of organizing centers may also play a part, as Petzelt (88) points out. In the NH_4OH activated eggs, the peak of the Ca^{2+}–ATPase activity normally found in the first part of the cell cycle is missing, a peak which he believes is related to the centriole cycle and which may involve a different pathway of tubulin polymerization. While it may be that the calcium pumps simply have not yet been triggered, due to low initial cytoplasmic calcium, the idea that normal centriole duplication is somehow related to calcium levels in the cytoplasm is a very interesting one. Evidence to support this idea comes from observations by Dustin *et al.* (28) that there is a tendency in cells treated with antimitotic agents to generate extra centrioles or form cilia, even in cells not normally possessing cilia. Other workers have also noted this effect (77, 83, 128). Filaments with a diameter of about 80 A are apparently another drug-resistant form of polymerized tubulin, first described by Robbins and Gonatas (106) in colchicine- and vinblastine-treated interphase HeLa cells. Thus, one might suppose that calcium, acting as an "antimitotic" agent, provides a large pool of available tubulin and shifts conditions to those that favor assembly into more stable, drug-resistant structures when microtubule polymerization is inhibited. In this regard, it should be recalled that tubulin polymerization is a biased bidirectional process, and, while polymerization into microtubules may be favored, the less favored direction may produce other structures *in vivo*.

The relationship of the centriole cycle to DNA synthesis was investigated by Bucher and Mazia (15), using mercaptoethanol, which blocks sea urchin mitosis and prevents the duplication of centers (73). They found that tritiated thymidine was incorporated into DNA during the mercaptoethanol block, regardless of whether or not the centers had already duplicated. The reciprocal of this experiment, the demonstration that centrioles could initiate new daughter centrioles in the absence of DNA synthesis, was reported by Rattner and Phillips (96). Using mouse L929 cell culture and arabinosyl cytosine as an inhibitor of DNA synthesis, they compared the fine structure events of normal mitosis with those of inhibited cells. Although procentrioles were associated with each of the two mature centrioles in the G_1 period, the normal events of prophase never occurred. The two centriole pairs did not migrate to opposite poles, nor did the daughter centrioles grow to their mature length as they normally do at prophase. Whether inhibition of centriole migration and elongation was a side effect of the continued presence of inhibitor is not known, but these results indicate that the S period may be the time in the cell cycle where nuclear events are coupled with cytoplasmic ones.

V. TRIGGERS AND TRIGGER WAVES

A. The Trigger for Calcium Release

If our assumptions concerning the role of calcium are correct, the entry of a G_2 cell into mitosis is initiated by the release of calcium that has been previously sequestered in an extensive cytoplasmic membrane system. This release, in turn, sets in motion a series of events culminating in completion of the mitotic process. If calcium is the trigger for mitosis at this particular stage, we must now ask what triggers the trigger.

Studies on the slime mold *Physarum polycephalum* may provide some clues. Both actin and myosin have been extracted from the plasmodium and are believed to be responsible for the shuttle streaming. Recently, calcium fluxes related to the streaming oscillations have been demonstrated with the use of injected aequorin (104) and by cytochemical localization of Ca^{2+} with the electron microscope (12). Sachsenmaier (110) has shown that this shuttle streaming stops during the synchronous nuclear division and suggested that a change in cytoplasmic calcium concentration may be responsible.

Since shuttle streaming in an extensive plasmodium is not usually synchronous throughout, some common, very rapidly diffusible cytoplasmic factor must be operating to bring about the synchronization. Gerson (44) has found that in *Physarum* there is a gradual approach during the cell cycle to a critical cytoplasmic pH at mitosis, and which may act as a trigger. If we can generalize about the nature of mitotic triggers, this would signal the entry of the plasmodium into antephase, characterized by a general release of sequestered calcium. This, in turn, would synchronize all the calcium pumps, lowering the cytoplasmic calcium below the threshold for streaming, and favoring the formation of microtubules in the many intranuclear spindles. While this is all very speculative, a critical pH is something that should be seriously considered as a mitotic trigger.

Durham (27), in a very stimulating review on control of actin and myosin in nonmuscle movements, cites numerous examples of calcium acting as a stimulator of cell proliferation. The situation is very confusing, however, when one considers the reciprocal relationship between Ca^{2+} and cAMP (reviewed by Rasmussen, *et al.* 95) and the several membrane-bounded compartments, i.e., mitochondria, sarcoplasmic reticulum, or microsomes, and the plasma membrane, which serve to sequester calcium, each in its own way. Many other factors, no doubt, play a role—for example, the oxidation–reduction cycle of glutathione (78), which appears to affect the switching on and off of the calcium pump, among other things. Although we have some hints, the question, "What triggers the trigger?" remains unanswered.

B. Trigger Waves: The Pattern of Calcium Release and Sequestering

In another study related to the calcium trigger, Endo *et al.* (29) showed that calcium itself could trigger the release of calcium from the sarcoplasmic reticulum of skinned muscle fibers. In this preparation, repeated contractions were observed in a medium containing appropriate concentrations of caffeine, Ca^{2+}, and chelating agent, suggesting that calcium release is a regenerative process, in which calcium itself causes the release of calcium from the reticulum, and this, in turn, stimulates the calcium pumps for its reloading.

In this case, two incompatable reactions are occurring: (1) release of calcium from the sarcoplasmic reticulum, and (2) reloading of calcium into the sarcoplasmic reticulum; but they can both occur in a homogeneous medium because they are separated in time. If, however, the first reaction were initiated at some point (the cell center, for example) and moved outward as a trigger wave of calcium release followed by the second reaction of calcium sequestering, both reactions could exist at the same time, since they would be separated in space. This is what happens in the fertilization reaction so beautifully demonstrated by Ridgway *et al.* (105) in *Medaka.* It is probably the basis of rhythmic peristaltic waves of contraction described by Schroeder (117) in various eggs, and it is very likely the pattern of calcium release and sequestering in the breakdown and formation of interphase and mitotic asters. The singular lack of success of many investigators to make chromosomes move in isolated mitotic apparatus may be due to the fact that the preparations are expected to operate in a homogeneous medium with unchanging ionic composition in time.

There are several advantages of such a pattern of spacially separated regions of calcium release and sequestering. First of all, it orients the newly polymerizing microtubules radially from the cell center. Second, the relatively short distance between the release and the sequestering permits a fairly high concentration of calcium to be released in a restricted region without affecting other parts of the cell. This can bring about the complete breakdown of any extraneous organizing centers and the suppression of unwanted structures that might arise from polymerization onto microtubule fragments. The result of such incomplete breakdown was demonstrated by Fuseler *et al.* (42) in cold-treated cells, which on warming had microtubules arising not only from the centers, but from more diffuse areas of the cytoplasm as well.

VI. A NEW MODEL FOR MITOSIS

A. Some Assumptions

If we now look at the problem of chromosome movements, the key factor appears to be the spacial and temporal control of cytoplasmic calcium ion concentration. This is important, regardless of whether the model assumes the microtubules alone are the force producers, or whether they have some help from actin–myosin interactions. For a start, however, let us build a model based on the structural changes observed during sea urchin mitosis, and what we know from our previous review about the chemical and physical characteristics of those structures. This allows us the following assumptions:

1. The microtubules are the force producers (but may have help).
2. Microtubules under physiological conditions show biased growth and breakdown; dimers are added only at the distal end, while breakdown involves, first, a structural deformation and an amorphous intermediate before the dimers are again released at the aster centers.
3. The control of microtubule polymerization is through regulation of cytoplasmic calcium ion concentration by a vesicular membrane system. Other ions and intermediates are probably also involved as a result of the calcium regulation.
4. The release and sequestering of calcium is initiated at the aster center and proceeds outward as a self-propagated trigger wave.
5. The trigger itself could be any one of a number of possible controls: pH, cyclic AMP, -SH, tubulin concentration, etc., which could build up to a critical level and set the process going.

B. How Does the Model Work?

Several explanations might be proposed for the microtubule deformation seen in the centrosphere. It could be an initial step in the breakdown of microtubules in the region where calcium concentration is again increasing. It is not uncommon to find deformed microtubules *in vitro* under conditions that marginally support polymerization. It could be that the membrane system, now releasing calcium and losing internal water, begins to contract and, thus, deforms the microtubules enmeshed within it. Another explanation is that the increasing calcium now allows the interaction of actin and myosin to form a contractile meshwork. Perhaps all these factors interact to provide a pulling force on the chromosomes through mechanical deformation of the microtubules, to provide an anchor against which this force can work, and finally, to permit the tubulin dimers to be released again for reuse.

Figure 7 shows how this system might work. The shaded areas represent zones of higher calcium ion concentration, which alternate with the zones where calcium is being actively sequestered. The boundaries of these zones move outward as self-propagating trigger waves from the cell center. (a) Interphase, with low cytoplasmic calcium and an extensive microtubule system. (b) A release of sequestered calcium moves outward from the cell center. The centrioles separate and interphase microtubules are depolymerized. (c) A new wave of calcium sequestering is begun, initiated now by two organizing centers instead of one. (d) As the asters grow at the periphery, calcium release begins at the aster centers. (e) Late anaphase or telophase, showing increased aster size. (f) In the daughter cells, sequestering is again initiated at the cell center to form the interphase aster, bringing the cycle back to (a).

With regard to the mechanism of chromosome movement, this system has two reactions, or forces, acting against each other: (a) Growth of microtubules by addition of dimers at their outer ends, where calcium is being sequestered, effectively *lengthening* the aster and spindle microtubules. (b) Deformation or bending of microtubules in the centrosphere prior to their depolymerization as a result of calcium release, effectively *shortening* them. Whether this deformation is a direct effect of calcium on the microtubules, or the effect of their being enmeshed in a nonmicrotubule contractile system, the result in either case is to pull the aster and spindle microtubules into the aster center.

The results of varying the two forces with respect to each other are obvious. When the two forces balance, there is a steady cycling of dimers through the system, with no net growth. This would be similar to Rebhun's (102) balanced reactions with inhibitors and augmentors. Chromosomes are pulled toward the aster centers when the growing ends of microtubules are blocked by attachment to the kinetochores, and force (b) predominates. Open-ended astral and spindle microtubules continue to grow as long as

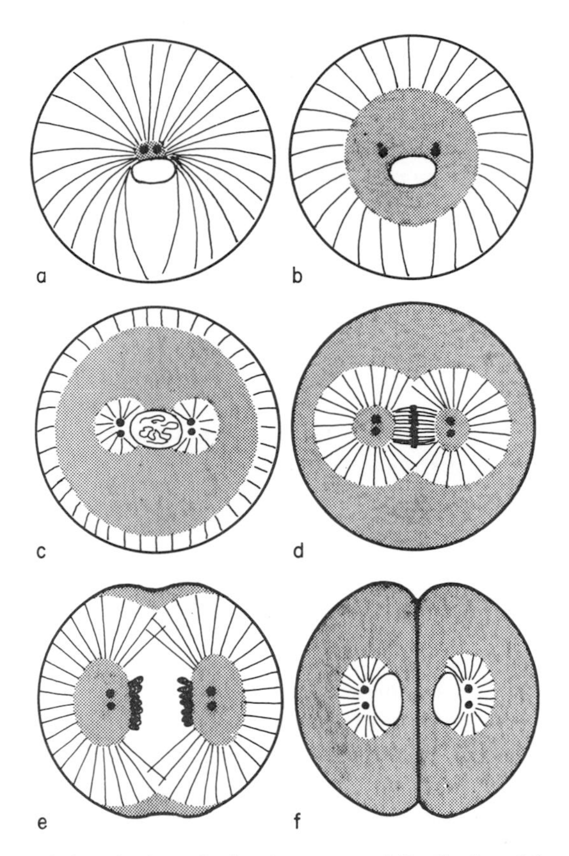

FIG. 7. Stages of the mitotic cycle showing proposed distribution of free cytoplasmic calcium (shaded) and sequestered calcium (unshaded). (a) Interphase; (b) "antephase" release of calcium; (c) early prophase; (d) metaphase; (e) late anaphase and telophase; (f) daughter cells forming new interphase asters.

force (a) predominates. The net result is that any one point on a microtubule is moving at a steady rate toward the aster center, even when the aster itself is growing in size.

C. What Questions Does the Model Answer?

This model could explain the following observations which have not been satisfactorily explained by currently proposed models:

1. How chromosomal fibers could shorten while astral and spindle microtubules elongate, without invoking the existence of two different kinds of microtubules.

2. The migration of particles toward the aster center. The pattern of colchicine-sensitive saltatory particle movement described by Rebhun and others, which suggests skiers hanging onto, or dropping off of a tow rope, is compatible with this model, and can be explained by adhesion or some kind of weak bonding to moving microtubules.

3. Microtubule interaction or cross-bridging between kinetochore microtubules. Rather than being force producers, such cross-bridges would be a structural necessity to allow a bundle of microtubules to pull together as a unit.

4. The interaction of the mitotic apparatus with the cell cortex during cytokinesis. If the membrane system associated with the mitotic apparatus acts to sequester calcium, the geometry of aster growth first brings the sequestering system in contact with the actin containing cortex at the polar regions, allowing a relaxation at the poles and contraction in the furrow.

VII. CONCLUSION

An attempt has been made to correlate the structural changes observed during the cell cycle with periodic changes known or suspected to occur in cytoplasmic calcium ion concentrations. Within this framework, a model for mitosis has been proposed that involves alternating triggered release of calcium and sequestering from the cytoplasm by an intracellular membrane system. Such triggers and trigger waves are neither elaborate nor without precedent in biologic systems. There are, of course, many examples among the more familiar excitable membranes of nerve and muscle, but even plants are known to have spontaneous electrical activity. For example, action potentials have recently been reported in the fungus *Neurospora crassa* (118, and references therein). The widespread occurrence of such membrane action potentials associated with ion permeability changes suggests that it may be an important means of transmitting information from the outside, as well as controlling the intracellular events of mitosis.

REFERENCES

1. Afzelius, B. A., *Z. Zellforsch. Mikrosk. Anat.* **45,** 660 (1957).
2. Allen, C., and Borisy, G. G., *J. Mol. Biol.* **90,** 381 (1974).
3. Andersson, R., Lundholm, L., Mohme-Lundholm, E., and Nilsson, D., *Adv. Cyclic Nuceotide Res.* **1,** 213 (1972).
4. Bajer, A., *Chromosoma* **25,** 249 (1968).
5. Bajer, A., *Cytobios* **8,** 139 (1973).
6. Bajer, A., and Molé-Bajer, J., *Chromosoma* **7,** 558 (1956).

7. Bajer, A., and Molé-Bajer, J., *Int. Rev. Cytol., Suppl.* **3,** 1 (1972).
8. Bloom, G., and Bryan, J., *J. Cell Biol.* **70,** 386a (1976).
9. Borisy, G. G., and Olmsted, J. B., *Science* **177,** 1196 (1972).
10. Borisy, G. G., Marcum, J. M., Olmsted, J. B., Murphey, D. B., and Johnson, K. A., *Ann. N.Y. Acad. Sci.* **253,** 107 (1975).
11. Borle, A. B., *J. Memb. Biol.* **16,** 221 (1974).
12. Braatz, R., *Cytobiologie* **12,** 74 (1975).
13. Brinkley, B. R., Fuller, G. M., and Highfield, D. P., *Proc. Natl. Acad. Sci. U.S.A.* **72,** 4981 (1975).
14. Bryan, J., *Am. Zool.* **15,** 649 (1975).
15. Bucher, N. L. R., and Mazia, D., *J. Biophys. Biochem. Cytol.* **7,** 651 (1960).
16. Bullock, T. H., *in* "Physiological Triggers and Discontinuous Rate Processes" (T. H. Bullock, ed.), p. 1. Am. Physiol. Soc., Washington, D.C., 1957.
17. Cohen, W. D., and Rebhun, L. I., *J. Cell Sci.* **6,** 159 (1970).
18. Condeelis, J. S., Taylor, D. L., Moore, P. L., and Allen, R. D., *Exp. Cell Res.* **101,** 134 (1976).
19. Conway, C. M., and Metz, C. B., *Cell Tissue Res.* **150,** 271 (1974).
20. Dales, S., *Proc. Natl. Acad. Sci. U.S.A.* **50,** 268 (1963).
21. Dales, S., *J. Cell Biol.* **52,** 748 (1972).
22. Dales, S., *Ann. N.Y. Acad. Sci.* **253,** 440 (1975).
23. Dales, S., Hsu, K., and Nagayama, A., *J. Cell Biol.* **59,** 643 (1973).
24. De Brabander, M., and Borgers, M., *J. Cell Sci.* **19,** 331 (1975).
25. Dietze, G., and Hepp, K. D., *Biophys. Biochem. Res. Commun.* **46,** 269 (1972).
26. DuPraw, E. J., *Dev. Biol.* **12,** 53 (1965).
27. Durham, A. C. H., *Cell* **2,** 123 (1974).
28. Dustin, P., Hubert, J.-P., and Flament-Durand, J., *Ann. N.Y. Acad. Sci.* **253,** 670 (1975).
29. Endo, M., Tanaka, M., and Ogawa, Y., *Nature* (*London*) **228,** 34 (1970).
30. Epel, D., Steinhardt, R., Humphreys, T., and Mazia, D., *Dev. Biol.* **40,** 245 (1974).
31. Ettienne, E. M., *J. Gen. Physiol.* **56,** 168 (1970).
32. Forer, A., *in* "Cell Cycle Controls" (G. M. Padilla, I. L. Cameron, and A. M. Zimmerman, eds.), p. 319. Academic Press, New York, 1974.
33. Forer, A., and Zimmerman, A. M., *J. Cell Sci.* **16,** 481 (1974).
34. Forer, A., and Zimmerman, A. M., *J. Cell Sci.* **20,** 309 (1976).
35. Forer, A., and Zimmerman, A. M., *J. Cell Sci.* **20,** 329 (1976).
36. Forer, A., Kalnins, V. I., and Zimmerman, A. M., *J. Cell Sci.* **22,** 115 (1976).
37. Franke, W. W., Stadler, J., and Krien, S., *Beitr. Pathol.* **146,** 289 (1972).
38. Freed, J. J., and Lebowitz, M. M., *J. Cell Biol.* **45,** 334 (1970).
39. Fuller, G. M., and Brinkley, B. R., *J. Cell Biol.* **63,** 106a (1974).
40. Fuller, G. M., Artus, C. S., and Ellison, J. J., *J. Cell Biol.* **70,** 68a (1976).
41. Fuller, G. M., Brinkley, B. R., and Boughter, J. M., *Science* **187,** 948 (1975).
42. Fuseler, J. W., Jones, J. E., Fuller, G. M., and Brinkley, B. R., *J. Cell Biol.* **70,** 54a (1976).
43. Gallin, J. I., and Rosenthal, A. S., *J. Cell Biol.* **62,** 594 (1974).
44. Gerson, D. F., this volume, Chapter 6.
45. Goldman, R. D., and Rebhun, L. I., *J. Cell Sci.* **4,** 179 (1969).
46. Gould, R. R., and Borisy, G. G., *J. Cell Biol.* **70,** 43a (1976).
47. Haga, T., Abe, T., and Kurokawa, M., *FEBS Lett.* **39,** 291 (1974).
48. Harris, P., *J. Biophys. Biochem. Cytol.* **11,** 419 (1961).
49. Harris, P., *J. Cell Biol.* **14,** 475 (1962).
50. Harris, P., *Exp. Cell Res.* **48,** 569 (1967).

51. Harris, P., *Exp. Cell Res.* **94,** 409 (1975).
52. Harris, P., *Exp. Cell Res.* **97,** 63 (1976).
53. Harris, P., *J. Cell Biol.* **70,** 112a (1976).
54. Hasselbach, W., *in* "Molecular Bioenergetics and Macromolecular Biochemistry" (H. H. Weber, ed.), p. 149. Springer-Verlag, Berlin and New York, 1972.
55. Hepler, P. K. *in* "Mechanism and Control of Cell Division" (T. Rost and E. M. Gifford, Jr., eds.), p. 212. Dowden, Hutchinson and Ross, Stroudsburg, Pennsylvania, 1977.
56. Hepler, P. K., and Jackson, W. T., *J. Cell Sci.* **5,** 727 (1969).
57. Hinkley, R., and Telser, A., *Exp. Cell Res.* **86,** 161 (1974).
58. Hiramoto, Y., *Exp. Cell Res.* **56,** 209 (1969).
59. Inoué, S., and Sato, H., *J. Gen. Physiol.* **50,** 259 (1967).
60. Ishikawa, H., Bischoff, R., and Holtzer, H., *J. Cell Biol.* **43,** 312 (1969).
61. Kane, R. E., *J. Cell Biol.* **12,** 47 (1962).
62. Kessel, R. G., *J. Ultrastruct. Res., Suppl.* **10,** 5 (1968).
63. Kiehart, D. P., and Inoué, S., *J. Cell Biol.* **70,** 230a (1976).
64. Kirschner, M., Suter, M., Weingarten, M., and Littman, D., *Ann. N.Y. Acad. Sci.* **253,** 90 (1975).
65. Lazarides, E., and Weber, K., *Proc. Natl. Acad. Sci. U.S.A.* **71,** 2268 (1974).
66. Longo, F. J., *J. Morphol.* **138,** 207 (1971).
67. Longo, F. J., and Anderson, E., *J. Cell Biol.* **39,** 339 (1968).
68. Mazia, D., *J. Cell. Comp. Physiol.* **10,** 291 (1937).
69. Mazia, D. *in* "The Cell. Vol. III. Meiosis and Mitosis" (J. Brachet and A. E. Mirsky, eds.), p. 77. Academic Press, New York, 1961.
70. Mazia, D., *Proc. Natl. Acad. Sci. U.S.A.* **71,** 690 (1974).
71. Mazia, D., and Ruby, A., *Exp. Cell Res.* **85,** 167 (1974).
72. Mazia, D., and Zimmerman, A. M., *Exp. Cell Res.* **15,** 138 (1958).
73. Mazia, D., Harris, P., and Bibring, T., *J. Biophys. Biochem. Cytol.* **7,** 1 (1960).
74. Mazia, D., Petzelt, C., Williams, R. O., and Meza, I., *Exp. Cell Res.* **70,** 325 (1972).
75. McIntosh, J. R., Cande, W. Z., and Snyder, J. A., *in* "Molecules and Cell Movement" (S. Inoué and R. E. Stephens, eds.), p. 31. Raven, New York, 1975.
76. McIntosh, J. R., Hepler, P. K., and Van Wie, D. G., *Nature* (*London*) **224,** 659 (1969).
77. Milhaud, M., and Pappas, G. D., *J. Cell Biol.* **37,** 599 (1968).
78. Nath, J., and Rebhun, L. I., *J. Cell Biol.* **68,** 440 (1976).
79. Nath, J., and Rebhun, L. I., *J. Cell Biol.* **70,** 43a (1976).
80. Nicklas, R. B., and Koch, C. A. *Chromosoma* **39,** 1 (1972).
81. Olmsted, J. B., and Borisy, G. G. *Annu. Rev. Biochem.* **42,** 507 (1973).
82. Olmsted, J. B., and Borisy, G. G., *Biochemistry* **14,** 2996 (1975).
83. Osborn, M. and Weber, K., *Proc. Natl. Acad. Sci. U.S.A.* **73,** 867 (1976).
84. Östergren, G., Molé-Bajer, J., and Bajer, A., *Ann. N.Y. Acad. Sci.* **90,** 381 (1960).
85. Petzelt, C., *Exp. Cell Res.* **70,** 333 (1972).
86. Petzelt, C., *Exp. Cell Res.* **74,** 156 (1972).
87. Petzelt, C., *Exp. Cell Res.* **86,** 404 (1974).
88. Petzelt, C., *Exp. Cell Res.* **102,** 200 (1976).
89. Petzelt, C., and von Ledebur-Villiger, M., *Exp. Cell Res.* **81,** 87 (1973).
90. Pollard, T. D., and Weihing, R. R., *Crit. Rev. Biochem.* **2,** 1 (1974).
91. Porter, K. R., and Bonneville, M. A., "Fine Structure of Cells and Tissues," 4th ed. Lea & Febiger, Philadelphia, Pennsylvania, 1973.
92. Porter, K. R., and Machado, R. D., *J. Biophys. Biochem. Cytol.* **7,** 167 (1960).
93. Raff, R. A., *Am. Zool.* **15,** 661 (1975).
94. Raff, R. A., Greenhouse, G., Gross, K. W., and Gross, P. R., *J. Cell Biol.* **50,** 516 (1971).

95. Rasmussen, H., Goodman, D. B. P., and Tenenhouse, A., *Crit. Rev. Biochem.* **1,** 95 (1972).
96. Rattner, J. B., and Phillips, S. G., *J. Cell Biol.* **57,** 359 (1973).
97. Rebhun, L. I., *in* "The Cell in Mitosis" (L. Levine, ed.), p. 67. Academic Press, New York, 1963.
98. Rebhun, L. I., *J. Gen. Physiol.* **50,** 223 (1967).
99. Rebhun, L. I., *Int. Rev. Cytol.* **32,** 93 (1972).
100. Rebhun, L. I., and Sander, G. F., *J. Cell Biol.* **34,** 859 (1967).
101. Rebhun, L. I., and Sawada, N., *Protoplasma* **68,** 1 (1969).
102. Rebhun, L. I., Jemiolo, D., Ivy, N., Mellon, M., and Nath, J., *Ann. N.Y. Acad. Sci.* **253,** 362 (1975).
103. Reinold, M. R., and Stockem, W., *Cytobiologie* **6,** 182 (1972).
104. Ridgway, E. B., and Durham, A. C. H., *J. Cell Biol.* **69,** 223 (1976).
105. Ridgway, E. B., Gilkey, J. C., and Jaffe, L. F. *Proc. Natl. Acad. Sci. U.S.A.* **74,** 623 (1977).
106. Robbins, E., and Gonatas, N. K., *J. Histochem. Cytochem.* **12,** 704 (1964).
107. Robblee, L. S., and Shepro, D., *J. Cell Biol.* **67,** 364a (1975).
108. Rose, B., and Lowenstein, W. R., *Science* **190,** 1204 (1975).
109. Rosenbaum, J. L., Binder, L. I., Granett, S., Dentler, W. L., Snell, W., Sloboda, R., and Haimo, L., *Ann. N.Y. Acad. Sci.* **253,** 147 (1975).
110. Sachsenmaier, W., and Hansen, K., *in* "Biological and Biochemical Oscillators" (B. Chance *et al.*, eds.), p. 429. Academic Press, New York, 1973.
111. Salmon, E. D., *J. Cell Biol.* **65,** 603 (1975).
112. Salmon, E. D., *J. Cell Biol.* **66,** 114 (1975).
113. Sanger, J. W., *Proc. Natl. Acad. Sci. U.S.A.* **72,** 1913 (1975).
114. Sanger, J. W., *Proc. Natl. Acad. Sci. U.S.A.* **72,** 2451 (1975).
115. Sato, H., Ellis, G. W., and Inoué, S., *J. Cell Biol.* **67,** 501 (1975).
116. Schliwa, M., *J. Cell Biol.* **70,** 527 (1976).
117. Schroeder, T., *in* "Molecules and Cell Movement" (S. Inoué and R. E. Stephens, eds.), p. 305. Raven, New York, 1975.
118. Slayman, C. L., Long, W. S., and Gradmann, D., *Biochim. Biophys. Acta* **426,** 732 (1976).
119. Sluder, G., *J. Cell Biol.* **70,** 75 (1976).
120. Snyder, J. A., and McIntosch, J. R., *J. Cell Biol.* **67,** 744 (1975).
121. Soifer, D., ed. "The Biology of Cytoplasmic Microtubules" Ann. N.Y. Acad. Sci. No. 253. N.Y. Acad. Sci., New York, 1975.
122. Steinhardt, R., and Epel, D., *Proc. Natl. Acad. Sci. U.S.A.* **71,** 1915 (1974).
123. Steinhardt, R., and Mazia, D., *Nature* (*London*) **241,** 400 (1973).
124. Stephens, R. E., *in* "Biological Macromolecules" (S. N. Timasheff and G. P. Fasman, eds.), Vol. 5A, p. 355. Dekker, New York, 1971.
125. Stephens, R. E., *Biol. Bull.* **142,** 145 (1972).
126. Stephens, R. E., *J. Cell Biol.* **57,** 133 (1973).
127. Stubblefield, E., and Brinkley, B. R., *J. Cell Biol.* **30,** 645 (1966).
128. Taylor, D. L., Moore, P. L., Condeelis, J. S., and Allen, R. D. *Exp. Cell Res.* **101,** 127 (1976).
129. Telzer, B. R., Moses, M. J., and Rosenbaum, J. L. *Proc. Natl. Acad. Sci. U.S.A.* **72,** 4023 (1975).
130. Verhey, C. A., and Moyer, F. H., *J. Exp. Zool.* **164,** 195 (1967).
131. Weber, A., *J. Gen. Physiol.* **52,** 760 (1968).
132. Weber, K., Bibring, T., and Osborn, M., *Exp. Cell Res.* **95,** 111 (1975).

133. Weber, K., Pollack, R. E., and Bibring, T., *Proc. Natl. Acad. Sci. U.S.A.* **72,** 459 (1975).
134. Weisenberg, R. C., *Science* **177,** 1104 (1972).
135. Weisenberg, R. C., and Rosenfeld, A. C., *J. Cell Biol.* **64,** 146 (1975).
136. Weisenberg, R. C., Borisy, G. G., and Taylor, E. W., *Biochemistry* **7,** 4466 (1968).
137. Wilson, E. G., "The Cell in Development and Heredity," 3rd ed. Macmillan, New York, 1925.
138. Wilson, L., and Bryan, J., *Adv. Cell Mol. Biol.* **3,** (1974).
139. Wilt, F., and Mazia, D., *Dev. Biol.* **37,** 422 (1974).
140. Wischnitzer, S., *Int. Rev. Cytol.* **27,** 65 (1970).
141. Yagi, K., Mase, R., Sakaibara, I., and Asai, H., *J. Biol. Chem.* **240,** 2448 (1965).

6

Intracellular pH and the Mitotic Cycle in *Physarum* and Mammalian Cells

Donald F. Gerson

I. INTRODUCTION

Water is the liquid medium for life; most metabolically and mitotically active cells contain approximately 80% water. One property of water is that it self-ionizes to produce hydrogen (H^+) and hydroxyl (OH^-) ions. The concentration of H^+ in pure water is approximately 10^{-7} *M*, which corresponds to pH 7.0. Ionizable biochemicals affect and are affected by the pH, as are the equilibrium concentrations of all biochemical reactions involving ionizable molecules. Since enzymes and other proteins have ionizable groups that determine the three-dimensional structure of the molecule, pH also can influence the rate of reaction for conversions involving neutral molecules. The pH of cells is, thus, of considerable biochemical and metabolic (possibly developmental) importance, and living organisms have developed a homeostatic control system that maintains intracellular pH near pH 7 over a wide range of external conditions. While intracellular pH is controlled, it is not constant, which raises questions concerning the quality of the homeostasis and the effect of changes in the intracellular pH on life processes. Few of these questions have been studied, partly because of the difficulties of measuring intracellular pH. This review discusses the relations between intracellular pH and the mitotic cycle.

II. MEASUREMENT OF INTRACELLULAR pH

Intracellular pH is difficult to measure. One needs somehow to probe the interior of the cell and to obtain an average measure of intracellular hydrogen ion concentration without allowing extracellular material to leak into and, often, without killing the cell. A wide variety of methods have been attempted; a few are relatively successful; these are described below. A review of intracellular pH measurement has been published recently (24).

A. Homogenization

The most obvious approach to the measurement of intracellular pH is to remove the cytoplasm from the cell and measure the pH of the expressed cell sap with an ordinary pH electrode. This technique works quite well, despite its simplicity. If dilution is necessary in order to obtain sufficient total volume, distilled water is the solvent of choice. Dilution of a buffered solution will not change its pH, since it is the ratio between the ionized and associated forms of the buffer that determines the pH in an ideal case. Cell sap is probably not an ideal buffer, but the simplicity and speed of this approach make it useful for preliminary experiments, and it has been used on plant cells (4, 39, 49), sea urchin eggs (54), muscle (48b), human erythrocytes (6), and bacteria (58).

There are many potential problems associated with this type of measurement. The most severe is contamination with buffering materials from the cell wall, growth medium, glycocalyx, extracellular void volume, or the like. Cell wall components of both plants and microbes are known to have ion exchange capabilities, and equilibration of these substances with cellular contents could result in significant pH changes in the homogenate. Cells lacking significant amounts of extracellular material may be less subject to this type of error. Consistent although relatively small differences between the results obtained by homogenization and the results obtained by other methods have been reported for plant cells (39) and red blood cells (11). Another possible source of error, which is applicable to all methods currently in use, is that if intracellular compartments (e.g., mitochondria) are at a sufficiently deviant pH, the value obtained will not truly reflect the pH of the cytoplasm. At this stage in the technology of intracellular pH measurement, there is no suitable method of accurately measuring the pH of the intracellular compartments of most cells, and, thus, this argument, while valid, is not constructive at this time. An ultimate difficulty with consideration of the compartmentalization of hydrogen ions is that subdivisions of the cellular volume rapidly approach a size at which the concept of pH as a property of bulk solutions becomes less meaningful. In addition, while intracellular membranes may be relatively impermeable to hydrogen ions or may actively control the flow of hydrogen ions (e.g., Mitchell hypothesis of ATP generation in mitochondria), it must be remembered that hydrogen ions are highly mobile (36×10^{-4} cm/second, 25°C) compared to other ions such as sodium (5×10^{-4} cm/second, 25°C), and that this relatively greater mobility only increases if water molecules are "bound": H^+ mobility in ice is orders of magnitude greater than that of K^+ or Na^+. Wiggins (99) has discussed the effect of water binding on intracellular pH.

B. Micro pH Electrodes

Micro pH electrodes have been widely used in the measurement of intracellular pH. In careful hands, they can yield useful and reproducible data. Two types of micro pH electrode are currently in use: pH electrodes utilizing hydrogen ion-sensitive glass membranes and those utilizing antimony films. Successful versions of both types have tip diameters in the order of 1 μm. All microelectrodes used to measure intracellular pH will give readings that must be corrected for the membrane potential. This can be done either by simultaneously inserting a reference electrode (KCl-filled micropipette attached to a calomel electrode) directly into the cell, or by making separate measurements of the membrane potential either on the same cells or on a similar population of cells.

Glass micro pH electrodes of various designs (Fig. 1) have been used to measure intracellular pH in a variety of tissues and organisms. The basic method of construction (Fig. 1A) is to produce a standard microelectrode with a tip diameter in the order of 1 μm (with a mechanical electrode puller)

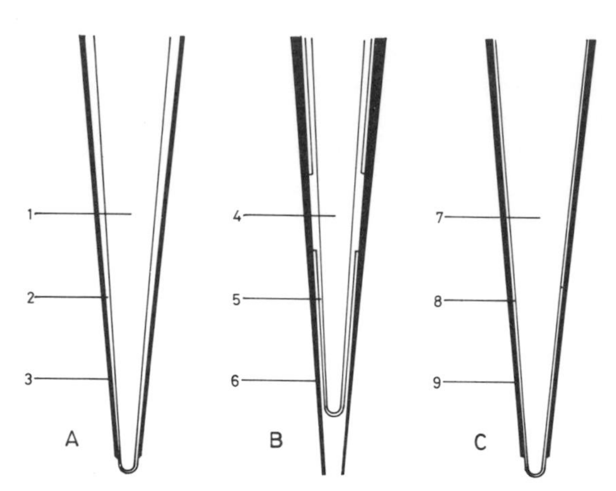

FIG. 1. Micro pH electrodes. (A) Conventional glass micro pH electrode: 1, lumen filled with distilled water, dilute buffer, or dilute acid; 2, glass pipette made from pH-sensitive glass and sealed at the tip; 3, insulation layer of glass glaze or other insulating material. (B) Recessed tip glass micro pH electrode: 4, lumen filled with dilute buffer; 5, glass pipette made from pH-sensitive glass, sealed at tip and bonded to external micro pipette; 6, external borosilicate glass micropipette [from Thomas (93)]. (C) Antimony micro pH electrode: 7, stainless steel or glass substrate; 8, vacuum deposited layer of antimony; 9, epoxy insulation. All drawings are schematic, tip diameters are in the order of 1 μm.

out of H^+-sensitive glass capillaries (e.g., Corning type 0150, Corning Glass Works, Medfield, Mass.). The tip is either sealed during the pulling operation or shortly thereafter with a microheater. Melting temporarily destroys the hydrogen ion sensitivity of this glass, and the electrodes must be rehydrated by several days of soaking in distilled water or dilute acid. As with any laboratory pH electrode, these electrodes follow the Nernst equation, which describes the potential generated across a membrane, which is permeable to only one ion and which has some finite concentration of that ion on both sides [Eq. (1)]:

$$\epsilon = \frac{RT}{\mathcal{F}} \ln \frac{[H_1^+]}{[H_2^+]} \tag{1}$$

In Eq. (1), ϵ is the potential, R the gas constant, T the absolute temperature, $\mathcal{F}$ Faraday's constant, and subscripts 1 and 2 refer to the sides of the membrane. At room temperature, there should be a change in potential of 58 mV for a tenfold change in hydrogen ion concentration on one side of the membrane; many handmade micro pH electrodes show lower slopes. For intracellular measurements, the shank of the electrode must be made impermeable to hydrogen ions to avoid measurements that are confounded by partial response to the pH of the bathing medium. Carter *et al.* (17) and Paillard (75) solved this problem by coating the capillary with a glass glaze (which is impermeable to hydrogen ions) before the electrode is pulled. Others have used shellac, epoxy resins, or wax and a large glass sheath (60). With all these insulation methods, significant questions remain concerning the size of the sensitive tip and the relative conductances of the tip and the shank. The overall conductance of these electrodes ranges from 10^{-11} to 10^{-9} mho (17; D. F. Gerson, unpublished data), which is close to the input impedance of many microelectrode amplifiers having field effect transistor input stages. Electrical noise is, thus, a problem. Further discussion of insulation problems may be found in Cohen and Iles (24).

Thomas (93) has developed a slightly different electrode (Fig. 1B) which simultaneously increases the conductance of the hydrogen ion-sensitive glass membrane and decreases the conductance of the sidewalls of the microelectrode. In this elegant design, a sealed tip, pH-sensitive glass microelectrode is inserted and fused into a standard borosilicate glass microelectrode. The sensitive tip then has considerable surface area exposed to a small volume enclosed by the borosilicate electrode. Fortunately, the small enclosed volume equilibrates rapidly with the solution in which the electrode is immersed. This is probably the most useful glass micro pH electrode design available at this time.

Antimony microelectrodes have been described by Bicher and Ohki (2). These electrodes are available commercially (Transydine General, Ann Arbor, Mich.) and consist of a thin vacuum-deposited film of antimony on either a glass micropipette or an etched stainless steel wire (Fig. 1C). Both designs have tip diameters in the order of 1 μm. Insulation is provided by a thin layer of epoxy resin. The conductance through the antimony to the solution is very much greater than leakage through the insulating film, and problems of interference by the pH of the bathing medium and noise resulting from a high-impedance source are greatly reduced. They give a linear response to pH over the range from pH 4 to 7, with approximately a 50-mV change in potential for each pH unit. These electrodes are subject to changes in calibration during use and should be recalibrated frequently. This may be due to corrosion of the antimony film. These electrodes have been used successfully on nerve cells (2), plants (7, 27, 39), a slime mold (40, 41), and mammalian cells (D. F. Gerson, unpublished data). In our experience, it is best to calibrate these electrodes in a homogenate of the cells that are to be measured and under conditions that will have a similar pO_2.

C. DMO Distribution

The weak acid, 5,5-dimethyloxazolidine-2,4-dione (DMO), distributes itself between the extracellular fluid and the cytoplasm, according to the pH of each of these compartments (95). The associated form of this molecule permeates through the cell membrane and achieves the same concentration within the cell as in the extracellular medium; it then ionizes according to the pH of each compartment. The technique involves equilibration of the cells with [^{14}C]DMO and an impermeable labeled compound, such as inulin, which can be used to measure the extracellular void volume. The method has been carefully reviewed by Cohen and Iles (24), and works well on animal tissues (83), plant tissues (39, 96), and bacteria (58).

D. Colorimetric and Fluorescent Indicators

Colorimetric determination of intracellular pH with indicators, such as bromthymol blue (pK_a = 6.8) or bromocresol green (pK_a = 4.5), has been tried many times, but usually the optical density of the intracellular dye is too low for accurate measurement, although it is possible to make a visual color estimation (29). Whole muscle intracellular pH has been measured with bromthymol blue by Herbst and Piontek (48a), and Chance (21) also used the absorbance of bromthymol blue as an indicator of intramitochondrial pH in preparations from rat liver, pigeon heart, and liver and kidney from guinea pigs, and was able to measure qualitative changes in pH due to calcium absorption and a variety of other factors. Yamaha (100) measured intracellular pH of plant cells with bromocresol green. Earlier, Chambers (20) and Needham and Needham (73) measured intracellular pH in marine eggs and other cells using a variety of pH-sensitive dyes.

Various fluorescent pH indicators have been used to measure qualitative changes in intracellular pH. The first class of these are fluorescent amines. Deamer, Prince, and Crofts (28) used the quenching of Atebrin and 9-aminoacridine to measure pH gradients across liposome membranes. They found that 9-aminoacridine was useful for pH gradients of 2–4 pH units within the range from pH 6 to 9, Atebrin was useful for smaller pH gradients, but exhibited nonideal behavior and anomalous fluorescence. Schuldiner, Rottenberg, and Avron (88) studied a variety of fluorescent amines as possible indicators of pH changes in chloroplasts. These amines are distributed between the medium and the chloroplasts in a manner analogous to the distribution of DMO. Uptake of these fluorescent amines by chloroplasts resulted in a quenching of total fluorescence, and uptake is dependent on the dissociation constant of the amine. The most useful fluorescent probe of intrachloroplastic pH changes was 9-aminoacridine.

The second class of fluorescent pH probes are coumarin derivatives. Grunhagen and Witt (45) used 6-hydroxycoumarin (umbelliferone) as an intrachloroplastic pH indicator. This molecule has a high quantum yield, a pK_a of approximately 8.0, and, being a natural plant product, has no influence on photosynthesis. The response of 6-hydroxy-4-methylcoumarin

(4-methylumbelliferone) to pH has been studied by Gomes (42) and Nakashima, Sousa, and Clapp (71). It also has a high quantum yield, has a pK_a of between 6.8 (43) and 7.5 (D. F. Gerson, K. Elisèvich, and E. Cannell, unpublished data). However, excessive binding limited its usefulness as an intramitochondrial pH indicator (21). We have tested the effect of 1% solutions of bovine serum albumin on the pH dependence of the fluorescence of both 6-hydroxy-4-methylcoumarin and 4-methylumbelliferone heptanoate and have found considerable reduction in the differential between fluorescence at low and high pH, thus severely limiting the usefulness of these dyes in measurements of changes in intracellular pH. Another coumarin derivative, 6,7-dihydroxy-4-methylcoumarin (4-methylesculetin), shows a classical increase in fluorescence with increasing pH and does not exhibit significant quenching by albumin or cell homogenate. This coumarin is easily absorbed by cells, does not appear to interfere with growth or mitosis, and has been used to measure changes in the intracellular pH of *Physarum polycephalum* (41). Fluorescent probes of intracellular pH have considerable advantages over micro pH electrodes or [^{14}C]DMO techniques in that they are noninterfering, have a rapid response to pH changes, and can be used for very long-term measurements.

III. EXTRACELLULAR pH AND INTRACELLULAR pH

Extracellular pH is used as a variable in many experiments and may affect many processes that occur in the extracellular space (23, 62), at the cell membrane (46, 57, 98), or within the cell (3, 36, 37). Changes in extracellular pH often result in changes in intracellular pH. The degree of this change is determined by an intracellular pH control mechanism which, while incompletely understood, depends on chemical buffering, metabolic adjustment, hydrogen ion transport across the cell membrane or into vacuoles, and organismal responses (lungs, kidneys) to alterations in blood composition and pH (22, 26, 80, 90).

A fundamental consideration in the discussion of the relation between extracellular pH and intracellular pH is whether the concentration of hydrogen ions within the cell is determined by electrochemical equilibrium or by active processes utilizing cellular energy to maintain some degree of homeostasis. If the Nernst potential for hydrogen ions (Eq. 1), calculated from measured extracellular and intracellular pH, is equal to the measured membrane potential (within experimental error), then the cell is in electrochemical equilibrium with respect to hydrogen ions. If these potentials are not equal, the hydrogen ion gradient may be maintained by the permeabilities of other ions (e.g., Donnan equilibrium), or it may be maintained by active transport of hydrogen ions. Both situations are known.

Kashket and Wong (58) have measured the intracellular pH as a function of extracellular pH in *Escherichia coli*. Over the range of extracellular pH from 5.0 to 9.0, intracellular pH (DMO or cell disruption) varies linearly with a slope of 0.6 from pH 5.5 to pH 8.0. Separate experiments demonstrate an important link between H^+ transport, a membrane-bound ATPase, and ATP production by estimation of the membrane potential with fluorescent probes in both *E. coli* and *Streptococcus lactis* (57, 64). The electrochemical gradient for hydrogen ions in *E. coli* is also linked to the active transport of β-galactosides (98). Harold *et al.* (47) has demonstrated that the pH gradient of *Streptococcus faecalis* is maintained by active processes. Measurements of intracellular pH and determinations of the electrochemical gradient in prokaryotes are sparse; however, acidophilic blue-green algae are known to live only above pH 4–5, and these organisms may be interesting candidates for studies of this type, especially in relation to mechanisms of intracellular pH control, which operate strictly at a cellular level. Eukaryotic algae are known to exist at hydrogen ion concentrations as high as pH 1.9, and these organisms may also be capable of extensive pH control (10). The same may be true for *Thiobacillus thiooxidans*, which can live in 1.0 *N* sulfuric acid. Along these lines, intracellular pH has been measured in *Thermoplasma acidophila* by Hsung and Haug (53). This is a thermophilic, acidophilic, mycoplasma-like organism which was grown at pH 2.0 and 50°C. Intracellular pH was measured with DMO and was found to be pH 6.6 ± 0.2. This value was constant over an extracellular pH range from 2.0 to 6.0. The intracellular pH of cells inhibited by 2,4-dinitrophenol, iodoacetate, or sodium azide, or killed by boiling for 5 hours was also 6.6. This indicates that control of intracellular pH in this organism may depend on an intracellular heat stable, nondiffusable ion-exchange material which maintains a hydrogen ion gradient of approximately 4.5 pH units by a mechanism akin to a passive Donnan equilibrium. Mitochondria are also thought to control intramitochondrial pH, in part, by a Donnan equilibrium (48).

Walker and Smith (96) studied both cytoplasmic and vacuolar pH in the green algae *Chara corallina* as a function of extracellular pH and found that, over the extracellular pH range from 5.3 to 7.0, cytoplasmic pH was nearly constant at pH 7.7. At slightly more basic external conditions up to pH 7.8, intracellular pH rises gently to pH 8.2. Over the external pH range from pH 6.2 to pH 7.8, vacuolar pH rises steadily from pH 5.3 to pH 6.5. Exchange of extracellular K^+ for intracellular H^+ and compartmentation of organic acid anions in the vacuole appear to be the most important mechanisms of pH regulation in both giant algal cells and higher plant cells

(80). Gerson (39) studied the effect of extracellular pH on intracellular pH in roots of *Phaseolus aureus* and found that intracellular pH is almost constant over the external pH range from 5.0 to 8.0, as is the membrane potential. At an extracellular pH of about 6.8, hydrogen ions are in electrochemical equilibrium across the membrane of these cells. Raven and Smith (80) and Davies (26) have discussed the control of intracellular pH in plants, and it is interesting to note that much of the control of intracellular pH involves transport processes. The transport of H^+ is often accomplished without the exchange of other ions to neutralize the net charge transfer, resulting in large changes in the membranes potential, which alters the electrochemical gradient of hydrogen and other ions, thus limiting the process (50).

Poole, Butler, and Waddell (77) studied the effects of extracellular pH and pCO_2 on intracellular pH (DMO) in Ehrlich ascites tumor cells. If the external pH was altered by adjustment of pCO_2, intracellular pH gradually changed from pH 6.4 to pH 7.2 as external pH varied from pH 6.3 to pH 7.4. Intracellular pH equaled extracellular pH at pH 7.1, although the membrane potential for these cells was not determined, and, thus, the point at which electrochemical equilibrium is reached cannot be determined. At a constant pCO_2 (35 mm Hg), intracellular pH was less variable and ranged from pH 6.8 to pH 7.2 as extracellular pH was changed from pH 6.4 to pH 7.6; the crossover point was pH 7.2. In both cases, the rate of change of intracellular pH with respect to extracellular pH increased with the hydrogen ion concentration. In the presence of phosphate buffers, at low pCO_2 (air), intracellular pH changed linearly from pH 6.6 to pH 7.4 over the same range of extracellular pH. The intracellular pH of this tumor line is not markedly acidic, as has been found for a variety of other tumors, albeit generally with less sophisticated methods (see below). The insensitivity of the intracellular pH of these cells to extracellular pH is notable, and may be related to the relative insensitivity of tumor cell growth to changes in extracellular pH (31). Carter, *et al* (17) observed that intracellular pH was a linear function of the membrane potential in voltage-clamped rat skeletal muscle, and intracellular pH varied from pH 8.0 to pH 3.0 as the membrane potential varied from +40 mV to −240 mV. Paillard (75) also measured the intracellular pH of rat skeletal muscle and found that intracellular pH was a linear function of extracellular pH at a relatively constant membrane potential; over an extracellular pH range from 7.0 to 8.0, intracellular pH ranged from pH 5.7 to pH 6.7. Paillard (75) also measured intracellular pH in crab muscle and found a linear relation; for external pH values between pH 3.5 and pH 8.5, intracellular pH rose from pH 5.9 to pH 7.0. The results of Paillard (75) and Carter *et al.* (17) may be explicable in terms of a Donnan equilibrium for H^+.

In the presence of lactate and a variety of buffers, Roos (83) obtained a nearly linear relation between intracellular pH and extracellular pH in rat muscle; as extracellular pH varied between 6.0 and 8.0, intracellular pH increased from pH 5.8 to pH 7.7, and, as with Ehrlich ascites cells, departure from linearity increased with increasing hydrogen ion concentration. Both Thomas (93) and Bicher and Ohki (2) measured the effect of short-term fluctuations of extracellular pH on intracellular pH of nerves from snail and squid, respectively. In each case, slight shifts in intracellular pH toward that of the extracellular pH were noted, with least control in the acid range. In sheep heart cells, Ellis and Thomas (34) measured intracellular pH with recessed tip micro pH electrodes (Fig. 1B), and in a bicarbonate buffer with variable pCO_2, intracellular pH was a linear function of extracellular pH (over the narrow extracellular pH range from 7.2 to 7.7, intracellular pH varied from pH 6.9 to pH 7.2). When bicarbonate was also varied, transient deviations of intracellular pH were observed.

In summary, intracellular pH of animal cells tends to be a nearly linear function of extracellular pH, with the greatest deviations from linearity occurring in relatively acidic extracellular fluids. One possible exception to this is noted by Poole, Butler, and Waddell (77); preliminary experiments with cells showing aerobic glycolysis gave an inverse relation between extracellular pH and intracellular pH. Considering the fundamental importance of these variables to cellular growth and metabolism, it is clear that insufficient study has been made of intracellular pH in relation to extracellular pH.

IV. EXTRACELLULAR pH AND THE GROWTH OF MAMMALIAN CELLS

Mammals maintain the hydrogen ion concentration of their blood with considerable constancy. In the human population, the pH of the blood is 7.40 ± 0.04. This narrow range and the stability of blood pH in individuals under normal circumstances reflects the quality of the pH control system of the body. On an intracellular level, and also in blood plasma, phosphate and, most importantly, bicarbonate ions provide simple chemical buffering against pH changes. Carbonic anhydrase, catalyzing the conversion of carbon dioxide to bicarbonate increases the dynamic responsiveness of the bicarbonate–carbon dioxide buffering system. As well, metabolic production and utilization of organic acids, such as lactate (malate in plants), adds considerable apparent buffering capacity.

On a tissue level, ion transport phenomena involving hydrogen ions allow the cell to control hydrogen efflux and influx usually by exchange mechanisms with alkali cations (Na^+, K^+), although, in certain cases, H^+ is pumped without charge balance by electrogenic pumps, which ultimately are controlled by the membrane potential. In many cases, the electrochemical potential gradient of H^+ ions across the cell membrane is not in equilibrium, implying that active transport of hydrogen ions occurs and is part of the pH control system. The hydrogen ion concentration in blood plasma is buffered not only by inorganic buffers, but also by the hemoglobin in red blood cells, and, to a lesser extent, by albumins and other plasma proteins. The buffering capacity of oxyhemoglobin is somewhat greater than that of deoxyhemoglobin, due to a decrease in the pK_a of hemoglobin, which occurs on oxygenation. Hydrogen ion concentration in blood plasma is also controlled by carbon dioxide exchange in the alveoli of the lungs and by the control of H^+ and HCO_3^- secretion and reabsorption by the kidney tubule. The pervasive and hierarchical nature of this control system results in excellent control of blood plasma pH and intracellular pH. The average intracellular pH of a human is 6.9 (65), although each organ tends to have its own distinct average intracellular pH.

The quality of pH control in a living organism strongly suggests that maintenance of a constant and optimal extracellular pH is of prime importance in any attempt to grow mammalian cells in tissue culture. With the use of nontoxic, nonvolatile, high capacity buffers (44), Eagle (30, 31) has been able to perform careful studies of the effect of extracellular pH on cell growth rate (protein), plating efficiency, and serum requirements of both normal and cancerous or virus-transformed cell lines. Normal human cell lines have a sharp pH optimum for growth (protein) on the average at pH 7.7; at approximately 0.2 pH units either side of the optimum, total cellular protein drops by 20%. The steepness of these nearly triangular curves of cell growth versus pH depends strongly on the serum concentration in the growth medium. At the optimal pH, cell growth is a linear function of serum concentration, while on either side of the optimum pH, cell growth is increasingly less sensitive to serum. Froehlich and Anastassiades (38) studied the effect of pH and serum concentration on the growth of human diploid fibroblasts. They also found that at the optimal pH for growth, cells were subject to greater stimulation by serum than were cells grown at a suboptimal pH. The pH that supported greatest growth did not depend on serum concentration, and was the same as that which maximized the likihood of clone formation following plating (31). This pH also optimized the number of cells per clone, demonstrating a close correlation between total cell protein and cell number. Cell lines from mice and rats have a similar response to pH, but the relative peak height at the optimum pH is less pronounced than that seen for human cells. These experiments show that the initiation of mitosis, and the continued increase in cell number to the point of density-dependent mitotic inhibition, requires a precise extracellular hydrogen ion concentration.

A dominant feature of the growth of mammalian cells is that growth and cell division are inhibited at high cell densities. In tissue cultures, this occurs with the formation of a confluent monolayer of cells, although the cell density of this monolayer is variable. For Chinese hamster lung fibroblasts grown in Eagle's minimal medium with 1% fetal calf serum, this limit is reached at approximately 5.0–5.5 contacts per cell, with an overall cell density of approximately 3×10^5 cells/cm^2 (52). In a three-dimensional tissue consisting of deformable, spherical cells, one would expect this to occur at or near to 13.6 intercellular contacts per cell, which is the average number of contacts found in the pith of *Eupatorium purpureum* and is the number of contacts found in the dense random packing of deformable spheres (67, 68, 89). Ceccarini and Eagle (18, 19) studied the effect of extracellular pH on the contact inhibition of growth of human diploid fibroblasts. At the pH that is optimal for growth, maximum cell densities (cells/cm^2) are 2–4 times that found when extracellular pH is allowed to vary (CO_2–HCO_3^- buffering). The stability of the pH of the growth medium was a more dominant factor controlling ultimate cell density than was nutrient supply or serum concentration. With bicarbonate buffering, maximal cell densities were between 1.2 and 1.5×10^5 cells (70–90 μg protein)/cm^2, while with well-buffered medium at the optimal pH, cell densities increased to 4–5 $\times$ 10^5 cells (270–300 μg protein)/cm^2. These higher values could not be further increased by refeeding. Lie, McKusick, and Neufeld (62) also concluded that the cell density at which contact inhibition occurred was a function of extracellular pH. Rubin (84, 85) studied the effects of extracellular pH on the growth and contact inhibition of chicken embryo fibroblasts in tissue culture. With bicarbonate buffering, pH changes in the medium were as much as 1.0 pH units by the termination of the experiment. In this system, growth increased with increasing pH over the range from pH 6.9 to pH 8.2. As in the work of Ceccarini and Eagle (18, 19), the effects of environmental pH became most prominent near confluency. The differences between these results and those of Ceccarini and Eagle (18, 19) and Eagle (31) may be due to the undifferentiated nature of embryo cells (see below), or to the much lower cell densities studied by Rubin. Both Taylor (92a) and Rubin (84) demonstrated that, in biocarbonate-buffered systems, hydrogen ion effects greatly exceed the effects of carbon dioxide or bicarbonate.

Cultured cancer or virus-transformed cells have a much less pronounced optimal pH than do normal cells, and generally have a more acid optimal pH for growth (18, 19, 31, 84). The average optimal pH for such cell lines was pH 7.2 (versus pH 7.7 for normal cells), and these cells produced more acid during growth than did normal cells (31). All cancerous or transformed cell lines grew more rapidly and to greater densities than normal cells when grown at the optimal pH; however, all cells eventually exhibited some degree of contact inhibition. It is striking that all cancer or transformed cell lines which have been studied have an abnormal response to pH, especially as the culture nears confluency. This observation covers a wide range of cell types, including a Rous sarcoma (84), HeLa, mouse multiple myeloma, rat glial tumor, and rat liver carcinoma cells (31). The broad pH range found for chick embryo cells by Rubin (84) may be due to some similarities between embryonic and neoplastic cellular growth control mechanisms.

Intercellular communication has been shown to be a factor controlling contact inhibition of growth and cell division by Loewenstein (63). Cancer cells lack the normal degree of intercellular communication and, of course, lack normal contact inhibition. Cancer cells also have an abnormal response to extracellular pH. In light of these and other findings, it has been suggested that extracellular pH either modifies the cell membrane, thus possibly modifying intercellular communication, or that hydrogen ions are the substance being transferred between communicating cells and that extracellular pH affects communication by altering pH gradients across the membrane, or by altering intracellular pH (12–14). If intercellular communication of hydrogen ions is important in contact inhibition, then there must be a mechanism linking hydrogen ions or intracellular pH to cell division (DNA synthesis). It should be noted that the optimal pH for growth in tissue culture (maximum final cell density) is the pH at which contact inhibition is least able to control growth, although, for most cell lines, contact inhibition eventually supercedes the effects of extracellular pH. Experiments have been performed (18) that demonstrate the early onset of contact inhibition at low cell densities following a shift to a suboptimal pH. The effects of environmental pH on the growth of Chinese hamster lung fibroblasts (V79-171b obtained from Dr. R. M. Sutherland, Ontario Cancer Foundation), have been studied with the help of Mr. K. Elisèvich, Mr. E. Cannell, and Mr. R. Howell. The cells were grown in Eagle's basal medium supplemented with 2.0 m*M* L-*glutamine, penicillin (100 IU/ml), strep*tomycin (100 μg/ml), and 15% fetal calf serum (GIBCO). The medium was buffered with 10 m*M* of each of three organic buffers: pipes, hepes, and tricine (30).

The buffering capacity of this mixture was measured and found to be 4 times that of the bicarbonate–carbon dioxide system alone, and was constant over the range from pH 6.8 to pH 8.0. The growth medium changed by less than 0.1 pH unit over the growth period. Incubation was carried out at 37°C in a humidified cabinet perfused with air plus 7% CO_2. Under these conditions, the midpoint of the logarithmic phase of growth occurred 90 hours after inoculation, and the maximum cell density (pH

7.40) was approximately 4×10^5 cells/cm². Cell counts (five plates, five counts per plate) were performed with a hemocytometer and a Biophysics Instruments model 6300A cell counter following 90 hours of growth at various hydrogen ion concentrations. A Corning model 101 digital pH meter with a Radiometer combination electrode was used to measure pH. When cultured under these conditions, Chinese hamster lung fibroblasts have a sharp growth maximum at pH 7.36 (Fig. 2).

Carbonic anhydrase plays an important role in the regulation of the balance between carbon dioxide and bicarbonate, and, thus, the pH of the

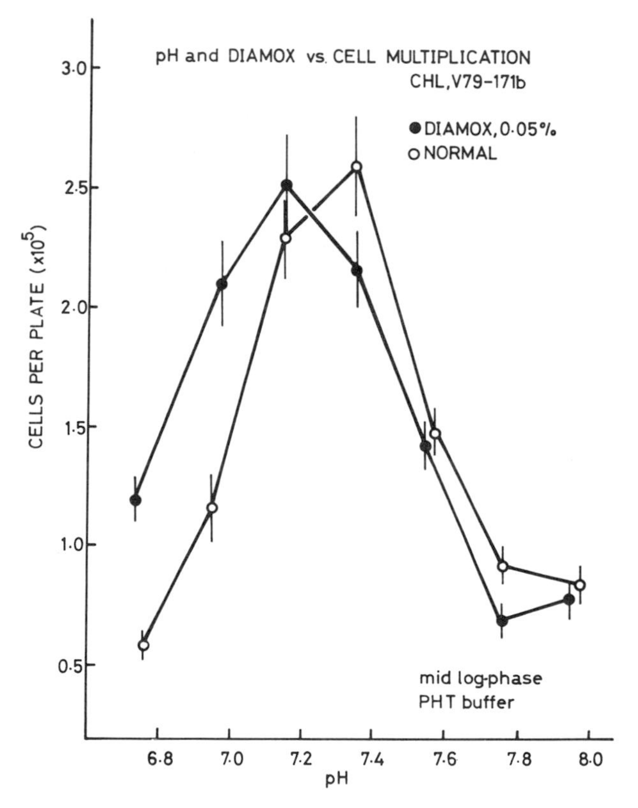

FIG. 2. Growth of Chinese hamster lung fibroblasts (V79-17lb) versus pH. Cells were grown in 6-cm diameter petri dishes at 37°C and 7% CO_2. Buffering was supplemented with 10 m*M* pipes, hepes, and tricine (PHT). Open circles give cell number per plate at 90 hours (mid-log-phase) under these growth conditions. Closed circles give cell number per plate at 90 hours in the presence of 0.50% (w/v) acetazolamide (Diamox). Error bars give standard error of the mean (SEM) of 25 counts. Values of pH are the average of initial and final pH; variation in pH was less than 0.1 pH unit over the growth period.

cell. In a preliminary attempt to investigate the relations between extracellular pH, intracellular pH, and cell growth (division), we attempted to modify intracellular pH with the carbonic anhydrase inhibitor acetazolamide (Diamox). The effect of acetazolamide on cultured cells presumably is to increase intracellular pH by decreasing the rate at which carbon dioxide reacts with water to form bicarbonate and hydrogen ions. The cell may be capable of a homeostatic response to acetazolamide, which would reduce the change in intracellular pH (e.g., H^+ transport), but the effect on intracellular acid–base balance would remain. It is known for a variety of systems that intracellular pH depends to some extent on extracellular pH (see above), and we postulated that the increased intracellular pH due to acetazolamide could be countered by a sufficiently acidic extracellular pH. This would have the effect of shifting the optimal extracellular pH for cell multiplication to a more acidic value. This hypothesis is confirmed in part by the results in Fig. 2, in which it is seen that cells grown with the addition of 0.05 mg/ml acetazolamide have a pH optimum at pH 7.16, which is 0.20 pH units more acidic than the normal pH optimum for growth. At this concentration, acetazolamide did not affect the mid-log-phase cell density (approximately 1×10^4 cells/cm^2), but it did slightly alter the shape of the curve relating growth to pH, increasing relative growth rates at hydrogen ion concentrations more acidic than the optimum. Measurement of the intracellular pH of these cells awaits further experimentation, and we must, as have others (31, 79), hope that someone will thoroughly study intracellular pH and cell division in relation to extracellular pH.

The results we have obtained demonstrate that the rate of cell multiplication is not only dependent on extracellular pH, but it is also dependent on intracellular pH or intracellular acid–base balance.

V. VARIATIONS OF INTRACELLULAR pH OVER THE CELL CYCLE

Interest in the intracellular pH of cells in relation to many variables, including cell division following fertilization, arose briefly between about 1925 and 1935, but the insensitive and tedious methods of injecting colorometric pH indicators into single cells failed to show large or reproducible changes. Needham and Needham (72, 73) studied both intracellular pH and intracellular redox potential with colorimetric indicators. The intracellular pH of the eggs of *Paracentrotus lividus* was measured both before and after fertilization and during early development, but within a range of $\pm$0.2–0.3 pH units, no changes occurred. Pandit and Chambers

(76), using similar methods, attempted to detect differences in intracellular pH between unfertilized and fertilized eggs of *Arbacia punctulata*, but the intracellular pH of the eggs stayed within the range of pH 6.8 ± 0.2. Recently, Johnson, Epel, and Paul (54) were able to detect pH changes in sea urchin eggs (*Strongylocentrotus purpuratus*) after fertilization, but the variation, while significant, was only from about pH 6.5 before fertilization to about pH 6.7 following fertilization. Yamaha (100) was able to estimate intracellular pH changes by observation of intracellular bromocresol green (pK_a = 4.5) during meiosis in two species of *Lilium, Tradescantia,* and *Reineckia.* In these plants, "cytoplasmic" (probably vacuolar) pH rose approximately 0.2 pH units at metaphase I, then dropped 0.2–0.3 pH units at telophase I, only to rise again by 0.1 pH unit at metaphase–anaphase II, followed by a decline of 0.2 pH units in the final tetrads.

Gerson and Burton (41) studied the changes of intracellular pH that occur over the mitotic cycle of *Physarum polycephalum. P. polycephalum* is an acellular slime mold that has naturally synchronous mitotic divisions every 8–12 hours, depending on the strain [ours had a very regular mitotic period of almost exactly 8 hours, and had been isolated locally by Holyer in 1964 (51)]. We developed a new fluorometric method of measuring changes in intracellular pH that involved incubation of the microplasmodia with a fluorescent pH indicator (6,7-dihydroxy–4-methylcoumarin) which has been used for pH measurements in wine (43). The plasmodium was fused and set up with indicator-free medium in the integrating sphere of a sensitive fluorometer in which it was exposed to 350 nm ultraviolet light that was chopped at a frequency of 400 Hz. The pH of the growth medium (5.2) is so low relative to the pK_a of the indicator that the only fluorescence emitted came from the plasmodium, and this was detected with a phase-lock amplifier operating at the same frequency and phase as the actinic UV light. In this way, changes in the order of 0.1 μV could be detected and distinguished from background noise. This apparatus detected a cyclical variation in the intracellular pH of *P. polycephalum* which had a period identical (±15 minutes) to the mitotic cycle, and which reached a maximum at the time of mitosis (Fig. 3A). Since the fluorescence changes depend on the size of the plasmodium and the intracellular concentration of fluorescent dye, it was necessary to make quantitative measurements of intracellular pH as well. These were made with antimony-on-glass or antimony-on-stainless steel micro pH electrodes. The average results (±SEM) for several cycles are given in Fig. 3B. Intracellular pH cycles between pH 5.9 and 6.0 at mid cycle and pH 6.5–6.7 at mitosis.

Unfortunately, little else is known of changes in intracellular pH over the cell cycle. Seemingly, however, it should be possible to use the [^{14}C]DMO

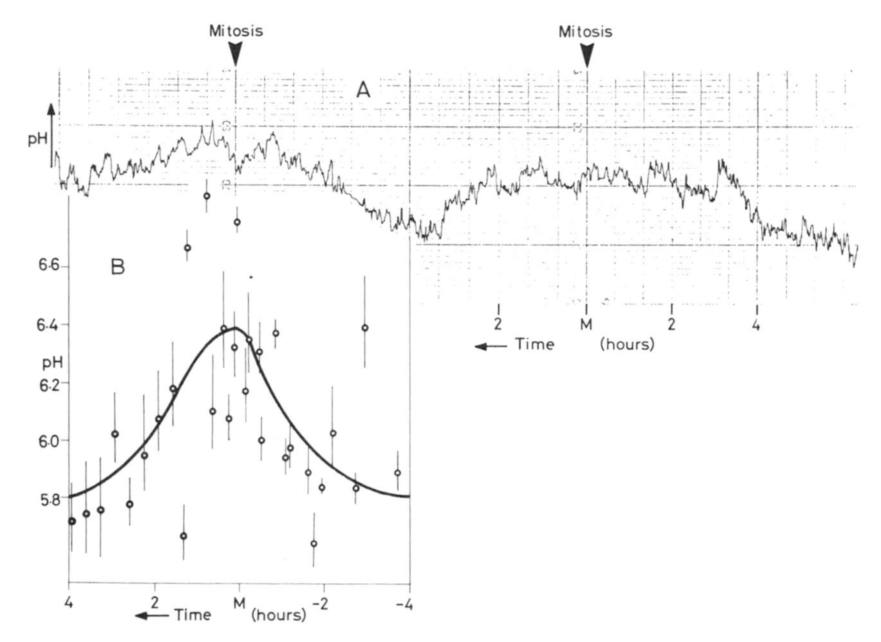

FIG. 3. Cyclical variation of intracellular pH in *Physarum polycephalum* over the mitotic cycle. (A) Qualitative detection of intracellular pH changes over an 18-hour period covering two mitotic divisions. Mitosis is indicated by arrows, time is from right to left, and one vertical division represents 0.1 μV. (B) Quantitative detection of intracellular pH changes over the cell cycle. Data are pooled from three mitotic cycles, each point ($\pm$ SEM) is the average of 10 measurements. Time is from right to left, mitosis is indicated by "M." Adapted from Gerson and Burton (41).

technique to make measurements on the intracellular pH of synchronous populations of cells such as *Tetrahymena* and *Euglena,* or during the first cycle following synchronization of tissue-cultured mammalian cells.

VI. MITOTIC CONTROL BY pH

The decision made during G_1 to proceed into S phase is almost always a decision to complete a mitotic cycle, since few cells that carry a second copy of the genome fail to divide. The following experiments suggest that H^+ is a mitotic control agent which acts during G_1 to stimulate the initiation of DNA synthesis. Every biochemical and physiological process is ultimately sensitive to pH. Eagle and his co-workers have shown that cell growth, cell division, and cell interaction depend on pH. On a biochemical scale, DNA synthesis, protein synthesis, and every enzyme catalyzed reaction depends on pH. Siskin and Kinosita (91) observed that extracellular pH changes affect the duration of G_1, as opposed to other phases of the cell cycle. The rate at which Chinese hamster Don cells proceed into the mitotic cycle from a colcemid block depends on pH. At pH 6.6, 30–40% of the cells are out of mitosis in 1 hour, while at pH 7.3 and pH 8.0, this percentage increases to 70% and 90–95%, respectively (74). Rubin (84, 86) found that both the rate of ^{14}C-thymidine uptake and the initiation of DNA synthesis increased with pH. Calothy *et al.* (16) studied the release of SV40 virus particles by mouse–monkey cell hybrids (Sendai virus fusion), and found 30,000 times as many virus particles are released at pH 8.4 as at pH 6.4, and that the response is exponential with respect to pH. Release of infective virus particles depends on replication of the viral genome, and, in monkey cells, the optimum is at pH 8.0. Fusion of mouse and monkey cells by Sendai virus is also optimal at pH 8.0. The similarly high pH optima for all of these phenomena suggests that pH affects the common limiting step, which is the initiation of DNA synthesis. This could also be true in *Physarum,* in which the rise in intracellular pH around the time of mitosis could be a stimulant to DNA synthesis, which follows within minutes of mitosis. F1 phosphorylated histone synthesis is important to chromosome condensation during prophase. Bradbury *et al.* (8, 9) have shown that there is a sharp rise in the phosphorylation of histone by F1 histone phosphokinase in *Physarum* about 1–2 hours prior to mitosis. At this time, intracellular pH is relatively high and is rising quickly (41). Intracellular pH at this time may assist or control ATP hydrolysis and the phosphorylation reaction, since the free energy of ATP hydrolysis is linearly dependent on pH (1). There are many other preparative steps prior to mitosis in *Physarum,* and it is possible to believe that the timing, pattern, and interplay of these processes could be affected by the slow cyclic variation in cytoplasmic pH that occurs over the cell cycle.

Johnson, Epel, and Paul (54) found that the intracellular pH of sea urchin eggs increases 0.2 pH units within 5 minutes after fertilization. This pH change results from the exchange of intracellular H^+ for extracellular Na^+ by a coupled transport system with a 1:1 stoichiometry (55). Replacement of extracellular Na^+ with choline or blockage of the transport system with amiloride prevents H^+ efflux and activation of the fertilized eggs. Immediately after fertilization, metabolic activity in the sea urchin egg increases dramatically, and, in 80–90 minutes, the first cell division occurs. The abrupt rise in intracellular pH following fertilization precedes and is required for metabolic activation and the initiation of cell division. They propose, as have Burton (13, 14) and Gerson and Burton (41), that an alteration of intracellular pH may be capable of encouraging a cell which is in G_1 (or G_0) to proceed with DNA synthesis. Stimulation of mitosis may occur as a result of either an increase or a decrease in intracellular pH (cf. Eagle's work). Recent results of C. R. Jones (Personal Communication) indicate that mitosis in *Euglena* is preceded by a decrease in intracellular pH. Of course, other cations (notably Ca^{2+}) are of great importance in these processes (82). However, the possibility that intracellular pH may alter the direction and the level of metabolic activity through alterations in the rates of a multitude of biochemical reactions and, thus, exert control over cell division is fascinating and deserves greater attention and study.

VII. ALTERED INTRACELLULAR pH AS A CORRELATE TO CANCER, A DISEASE OF THE CELL CYCLE

A. Acidosis and Cancer

Measurements of the pH of cancerous tissues have been made sporadically over the last 40 years, and show that cancer tissue is generally more acidic than normal tissue. Kahler and Robertson (56) inserted small pH electrodes into normal liver and hepatic tumors of rats and mice. Normal rat liver was at pH 7.4, while the hepatoma was at pH 7.0, mouse hepatoma was at pH 6.7. Meyer *et al.*, (69) measured the pH of cancerous human tissues by inserting pH electrodes into nonvascularized portions of surgically excised specimens. Malignant tissues were 0.5 pH units more acidic than normal tissues from the same area of the patient. Naeslund and Swenson (70) induced tumors in mice with methylcholanthrene and measured tissue pH with small pointed glass pH electrodes. Tumor tissue was approximately 0.6 pH units more acidic than normal tissue (liver). Eden, Haines, and Kahler (33) measured the tissue pH of eight different transplantable rat tumors and found that the average tumor was at pH 7.0, compared to pH 7.4 in normal tissues. Block (5) reported significant lactic acidosis in some cancer patients with uncontrolled tumor growth or during relapse following anti-tumor therapy. Kuhbock and Kummer (61) observed metabolic acidosis in 31 patients with severe cancer. Of course, none of these are true measures of intracellular pH, but they are indications of altered acid–base balance in tumor tissues. The intracellular pH of Ehrlich ascites tumor cells measured with DMO is 7.2 at physiological conditions (77, 78). The acidosis that accompanies cancerous growth is most probably associated with Warburg's well-known observation that cancer cells exhibit aerobic glycolysis, with the production of lactic acid (97). Racker (79) has expanded Warburg's observation and theory on the origin of cancer cells (97) to include not only alterations in glycolytic control but also alterations in all bioenergetic control mechanisms (see below).

B. Racker's Hypothesis

Racker (79) proposed that cancerous tissues may suffer from "a persistent alteration in the intracellular pH." Thus, any carcinogenic agent (virus, chemical, or electromagnetic radiation) may induce permanent alterations in the ability of a cell to maintain normal control over intracellular pH, and abnormal intracellular pH upsets bioenergetic control mechanisms. Many cells will die under these circumstances, but some will live and proliferate. Cells that survive and form cancerous tissues usually lack proper control of oxidation (both glycolytic and oxidative), which results in diminished ATP concentrations and high aerobic glycolsis. A key enzyme in glycolytic regulation is phosphofructokinase, which is inhibited by high ATP concentrations. Increased extracellular pH results in increased phosphofructokinase activity and increased glycolysis (37), while acidification of the intracellular pH (e.g., by increased aerobic glycolysis) increases the allosteric inhibition of this enzyme by ATP (79). Various factors, such as increased ATPase activity, increased ADP concentration, phosphate uptake, or alterations in H^+ transport, could result in intracellular acidification. In this regard, it is interesting to recall Eagle's (31) observation that all cancer and virus-transformed cells have a very broad pH optimum for growth, indicating an insensitivity to pH which would allow growth, albeit abnormal, at an altered intracellular pH. Unfortunately, thorough tests of Racker's interesting hypothesis have not been made. We already know that the intracellular pH of cancer cells is often abnormal (see above), but we do not know about relations between intracellular pH, aerobic glycolysis, and the rate of proliferation. At least two approaches are possible from a pharmacologic viewpoint: to search for glycolytic inhibitors, as Racker (79) has suggested, or to look for drugs that will correct the situation by alteration of the intracellular pH. Our experiments have indicated that acetazolamide (or any other carbonic anhydrase inhibitor) or chemicals such as ammonium chloride, which change pH through physiological response, may be useful in the second approach. An additional problem which must be recognized is that the pH change may be very slight in some cases as a result of the various "buffering" processes available to organisms, but a persistent alteration of intracellular acid–base balance (i.e., concentration of one or more of the intracellular buffers, the activity of H^+ pumps, membrane potential, etc.) could result in a significant disturbance of the metabolic control system. In addition, certain physiological problems, such as emphysema, acclimatization to altitude, achlorhydria, or altered salt balance, could ultimately result in intracellular pH changes that would affect the rates of cellular proliferation (13, 14).

Two studies have been made which support Racker's hypothesis, although they were not designed to test it. Flaks, Hamilton, and Clayson (35) induced bladder tumors in mice with 4-ethylsulfonylnaphthalene-1-sulfonamide (ENS). Addition of 0.01% ENS to the diet alkalinized the urine from pH 6.2 to pH 7.2. Addition of 1% ammonium chloride to the drinking water did not affect the pH of the urine of control animals, but acidified the urine of ENS fed mice to pH 6.5. Acidification of the urine with ammonium chloride completely eliminated bladder tumors from mice fed with ENS. In this study, a persistent change in pH resulted in carcinogenesis, and a return to normal pH prevented carcinogenesis. Kitagawa and Kuroiwa (59) studied azo dye carcinogenesis, using a carcinogen and an inactive isomer. The carcinogen, 3′-methyl-4-dimethylaminoazobenzene (3′-Me-DAB), produced a chronic reduction in the intracellular pH (DMO) of the liver, while the noncarcinogenetic isomer, 2-methyl-4-dimethylaminoazobenzene (2-Me-DAB), did not alter intracellular pH. Blood pH of rats fed 3′-Me-DAB was lower than that of rats fed 2-Me-DAB. Normal liver had an intracellular pH of 7.4, the liver of 3′-Me-DAB fed rats had a pH of 6.7, and the resulting cancerous nodules had a pH of 7.0. A single dose of carcinogen produced a significant ($p < 0.05$) decrease in both blood and liver pH. These results demonstrate a causal relation between carcinogenesis and pH change; however, we do not know whether the carcinogen caused the pH to change, which resulted in cancerous growth, or whether the carcinogen caused cancerous growth which resulted in a pH change.

There is some evidence that pH changes can result in mutation. Shimada and Ingalls (89a) observed chromosomal mutations following changes in extracellular pH. Anomolous chromosomes and polyploidy were present in cultured lymphocytes grown at hydrogen ion concentrations outside the physiological range. Aneuploidy was minimal at pH 7.4–8.0, but increased dramatically outside this range. Endoreduplication was especially common below pH 7.2. These authors suggest that local variations in pH within an organism may result from microbiological (or chemical) action and that this may lead to mutation, and, if the mutation is relatively stable, to cancer.

C. Burton's Altitude Effect

My colleague, Prof. Alan Burton, became interested in the effect of pH on cancer by way of his theory that mitotic control in tissues may be effected by asynchronous intracellular oscillations which are communicated between neighboring cells (12–15). Loewenstein has shown that normal cells communicate through tight junctions, while cancer cells do not (63). The theory predicts contact inhibition quite well (52), but it leaves us wondering what "key" chemical undergoes intracellular oscillations and is involved in intercellular communication. Hydrogen ions are small and can pass through tight junctions, and can have general effects on the metabolism and growth of cells. They are, thus, excellent candidates for the key substance. Burton (13) realized that nature had performed an experiment for us which indirectly tests the effect of pH on cancer growth.

At sufficiently high altitudes, anoxia stimulates the carotid body, which in turn increases the rate of breathing. This results in alleviation of anoxia, but causes excessive loss of respiratory CO_2, which results in alkalosis, or altitude sickness. After acclimatization, blood pH returns to normal, but acid–base relations are abnormal. Thus, a study of age-specific cancer incidence or death rates versus altitude could provide circumstantial evidence correlating pH and acid–base relations to cancer. The six countries of highest and lowest population-weighted mean altitude were compared for age-specific cancer rates in 5-year age intervals, using data from the World Health Organization. At ages less than about 60 years, the curves for both incidence and death rates climb almost exponentially, but, at age 60, the curve for high-altitude locations begins to level off, while that for low altitudes continues to climb. For a total of 16 countries, the correlation between altitude and cancer incidence for persons older than 60 years was negative ($r = -0.6$), and significant at the 95–99% level. This is true for both males and females. The distribution of types (anatomical location) of cancer did not change with altitude. In a separate study of leukemia incidence in the counties of the United States, Eckhoff *et al.* (32) found a negative correlation with altitude (while they were looking for a positive one due to decreased absorption of radiation by the atmosphere). In both studies, the altitude effect begins to occur at about 500 meters, which is approximately the altitude at which anoxia begins.

The correlation of cancer incidence with altitude is both remarkable and unexpected. It does not prove that cancer is related to or caused by an imbalance in pH or acid–base relations, but, together with the other information discussed above and by Burton (13, 14), it indicates that alterations in intracellular pH or acid–base balance may be of therapeutic value. The altitude effect indicates that alkalinization may prevent carcinogenesis; von Ardenne (94) has used hyperacidification as part of an anticancer chemotherapeutic method.

D. Cancer Therapy and Detection

Clinical treatment of cancer requires selective and virtually total destruction of cancer cells, with minimal damage of adjacent healthy tissues.

Aerobic glycolysis and the resulting disturbance of tissue pH and acid–base balance offers an approach to selectivity that has been utilized successfully by von Ardenne (94). Glycolyzing cells become hyperacidic when supplied with high concentrations of glucose (33, 56, 70, 81, 94). Glucose feeding or injection can result in differences between normal and cancer cells of 1 pH unit or more. Normal cells have an intracellular pH of about 7.4, while, under normal circumstances, cancer cells have a pH of about 7.0; with glucose feeding (400 μg/ml cells), cancer tissues can achieve pH 5.8–6.0. A tenfold difference in hydrogen ion concentration between normal and cancer cells offers the potential of high selectivity in the effects of drugs or other treatments. Naeslund and Swenson (70) attempted to utilize this difference with a drug that was insoluble at about pH 6.5, but soluble at pH 7.4, thereby precipitating and concentrating the drug selectively in cancer cells. Von Ardenne (94) has developed a gentle method of stimulating autolysis in cancer cells following hyperacidification with glucose. These clinical methods are promising and demonstrate the benefits of exaggerating, rather than normalizing, the unusual glycolytic and acid–base relations that are found in cancer cells.

An exciting cancer detection method has been devised and developed by Russell (87). Russell has found that the urine of cancer patients contains abnormally high quantities of the polyamines putrescene, spermidine, and spermine, and that that these levels drop after surgical removal of the tumor or successful chemotherapy. For example, the spermidine content of normal urine is about 1.5 mg/24 hours, while the average for various cancers and tumors is about 5.0 mg/24 hours; spermidine also increases in the urine of cancer patients by a factor of 10 (66). At first sight, this seems to have little to do with intracellular pH; however, plant physiologists and microbiologists have found that polyamines are produced by cells that are subject to conditions which tend to cause intracellular acidification. Potassium deficiency, for example, causes many plant roots to accumulate intracellular H^+, which cannot be exchanged for extracellular K^+; exposure to extracellular acid also stimulates polyamine synthesis, and it has been proposed that these basic amines are produced to titrate the cell to a more normal intracellular pH. Polyamines are also potent stimulators of nucleic acid synthesis (25, 92). Thus, polyamines could link intracellular pH to DNA synthesis and cell division; however, we know of no literature directly pertaining to this hypothesis.

VIII. CONCLUSIONS

Both intracellular and extracellular pH exert control over the mitotic cycle:

1. Glycolysis can be regulated by the effects of pH on the allosteric inhibition of phosphofructokinase, which is a key enzyme in the control of glycolytic metabolism.

2. The initiation of DNA synthesis is stimulated by increased pH in tissue culture of mammalian and chicken cells, and in viral replication. In *Physarum,* the initiation of DNA synthesis occurs at a time when the natural cycle of intracellular pH is near its maximum.

3. Both the rate of mitosis and the degree of contact inhibition of tissue cultures of mammalian cells show a sharp optimum with respect to extracellular pH. Manipulation of intracellular pH can result in a shift in the optimal extracellular pH for mitosis in cultured mammalian cells. In addition, the rise in intracellular pH of sea urchin eggs following fertilization may stimulate the first mitotic division of the zygote.

5. Hydrogen ion concentration anomalies have an important, but poorly studied, link to cancer, which is a mitotic disease. The pH of cancer cells is usually more acidic than normal; the growth of cancer cells in tissue culture is less dependent on pH than that of normal cells; alkalinization caused by high altitude is correlated with reduced cancer incidence. Many cancer cells produce excessive quantities of polyamines, and this may be a homeostatic response to intracellular acidosis.

ACKNOWLEDGMENTS

This study was initiated under a grant from the Cancer Research Fund, Faculty of Medicine, University of Western Ontario. I would like to acknowledge the laboratory assistance of O. Dawydiak, K. Elisèvich, E. Cannell, and R. Howell. The discussions and encouragement of my colleague, Dr. Alan C. Burton, have been invaluable in the completion of this work.

REFERENCES

1. Alberty, R. A., *J. Biol. Chem.* **244,** 3290 (1969).
2. Bicher, H. I., and Ohki, S., *Biochim. Biophys. Acta* **255,** 900 (1972).
3. Birmingham, M. K., and Elliott, K. A. C., *J. Biol. Chem.* **181,** 73 (1951).
4. Blevins, D. G., Hiatt, A. J., and Lowe, R. H., *Plant Physiol.* **54,** 82 (1974).
5. Block, J. B., *Ann. N.Y. Acad. Sci.* **230,** 94 (1974).
6. Bone, J. M., Verth, A., and Lambie, A. T. *Proc. Eur. Soc. Clin. Invest.* **9,** 24 (1975).
7. Bowling, D. J. F., *J. Exp. Bot.* **24,** 1041 (1973).
8. Bradbury, E. M., Inglis, R. J., and Matthews, H. R. *Nature* (*London*) **247,** 257 (1974).
9. Bradbury, E. M., Inglis, R. J., Matthews, H. R. and Langan, T. A., *Nature* (*London*) **249,** 553 (1974).
10. Brock, T. D., *Science* **179,** 480 (1973).
11. Bromberg, P. A., Theodore, J., Robin, E., and Jensen, W. *J. Lab. Clin. Med.* **66,** 464 (1965).

12. Burton, A. C., *Perspect. Biol. Med.* **14,** 301 (1971).
13. Burton, A. C., *Eur. J. Cancer* **11,** 365 (1975).
14. Burton, A. C., *Collect. Pap. Annu. Symp. Fundam. Cancer Res.* **28,** 249 (1975).
15. Burton, A. C., and Canham, P. B., *J. Theor. Biol.* **39,** 555 (1973).
16. Calothy, G., Croce, C., Defendi, V., Kaprowski, H., and Eagle, H., *Proc. Natl. Acad. Sci. U.S.A.* **70,** 366 (1973).
17. Carter, N. W., Rector, F. C., Campion, D. S., and Seldin, D. W., *J. Clin. Invest.* **46,** 920 (1967).
18. Ceccarini, C., and Eagle, H., *Nature* (*London*) New Biol. **233,** 271 (1971).
19. Ceccarini, C., and Eagle, H., *Proc. Natl. Acad. Sci. U.S.A.* **68,** 229 (1971).
20. Chambers, R., *J. Cell. Comp. Physiol.* **1,** 65 (1932).
21. Chance, B., *in* "Biochemistry of Mitochondria" (E. C. Slater, Z. Kaniuga, and L. Wojtczak, eds.), p. 93. Academic Press, New York, 1967.
22. Christensen, H. N., "Neutrality Control in the Living Organism." Saunders, Philadelphia, Pennsylvania, 1971.
23. Cleland, R., *Proc. Natl. Acad. Sci. U.S.A.* **70,** 3092 (1973).
24. Cohen, R. D., and Iles, R. A., *Crit. Rev. Clin. Lab. Sci.* **6,** 101 (1975).
25. Cohen, S. S. "Introduction to the Polyamines." Prentice-Hall, Englewood Cliffs, New Jersey, 1971.
26. Davies, D. D., *Symp. Soc. Exp. Biol.* **27,** 513 (1973).
27. Davis, R. F., *in* "Membrane Transport in Plants" (U. Zimmerman and J. Dainty, eds.), p. 197. Springer-Verlag, Berlin and New York, 1974.
28. Deamer, D. W., Prince, R. C., and Crofts, A. R., *Biochim. Biophys. Acta* **274,** 323 (1972).
29. Drawert, H., *Protoplasmologia* **2,** D3, 516 (1968).
30. Eagle, H., *Science* **174,** 500 (1971).
31. Eagle, H., *J. Cell. Physiol.* **82,** 1 (1973).
32. Eckhoff, N. D., Shultis, J. K., Clark, R. W., and Ramer, E. R., *Health Phys.* **27,** 377 (1974).
33. Eden, M., Haines, B., and Kahler, H., *J. Natl. Cancer Inst.* **16,** 541 (1955).
34. Ellis, D., and Thomas, R. C. *Nature* (*London*) **262,** 224 (1976).
35. Flaks, A., Hamilton, J. M., and Clayson, D. B., *J. Natl. Cancer Inst.* **51,** 2007 (1973).
36. Fodge, D. W., and Rubin, H., *J. Cell. Physiol.* **85,** 635 (1975).
37. Fodge, D. W., and Rubin, H., *J. Cell. Physiol.* **86,** 453 (1975).
38. Froehlich, J. E., and Anastassiades, T. P., *J. Cell. Physiol.* **84,** 253 (1974).
39. Gerson, D. F., *Plant Physiol.* **51,** 44s (1973).
40. Gerson, D. F., and Burton, A. C., *Can. Fed. Biol. Soc.* **18,** 12 (1975).
41. Gerson, D. F., and Burton, A. C., *J. Cell. Physiol.* **91,** 297 (1977).
42. Gomes, D. J. S., *Rev. Port. Farm.* **21,** 54 (1971).
43. Gomes, D. J. S., *Rev. Port. Farm.* **21,** 245 (1971).
44. Good, N., Winget, G., Winter, W., Connolly, T., Izawa, S., and Singh, R., *Biochemistry* **5,** 467 (1966).
45. Grunhagen, H. H., and Witt, H. T., *Z. Naturforsch. Teil B* **25,** 373 (1970).
46. Hagiwara, S., Gruener, R., Hayashi, H. Sakata, H., and Grinnell, A., *J. Gen. Physiol.* **52,** 773 (1968).
47. Harold, F. M., Pavlasova, E., and Baarda, J. R., *Biochim. Biophys. Acta* **196,** 235 (1970).
48. Harris, E. J., and Bangham, J. A., *J. Membr. Biol.* **9,** 141 (1972).
48a. Herbst, M., and Piontek, P., *Pfluegers Arch.* **335,** 213 (1972).
48b. Hermansen, L., and Osnes, J. B., *J. Appl. Physiol.* **32,** 304 (1972).
49. Hiatt, A. J., *Plant Physiol.* **42,** 294 (1967).

50. Higinbotham, N., *Bot. Rev.* **39,** 15 (1973).
51. Holyer, N., M.Sc. Thesis, University of Western Ontario, London, Canada (1964).
52. Howell, R. L., M.Sc. Thesis, University of Western Ontario, London, Canada (1975).
53. Hsung, J. C., and Haug, A., *Biochim. Biophys. Acta* **389,** 477 (1975).
54. Johnson, J. D., Epel, D., and Paul, M., *Nature* (*London*) **262,** 661 (1976).
55. Johnson, J. D., and Epel, D., *J. Cell Biol.* **70,** 382a (1976).
56. Kahler, H., and Robertson, W., *J. Natl. Cancer Inst.* **3,** 495 (1943).
57. Kashket, E. R., and Wilson, T. H., *Biochem. Biophys. Res. Commun.* **58,** 879 (1974).
58. Kashket, E. R., and Wong, P. T. S., *Biochim. Biophys. Acta* **193,** 212 (1969).
59. Kitagawa, Y., and Kuroiwa, Y., *Life Sci.* **18,** 441 (1976).
60. Kleinzeller, A., Kostyuk, P. G., Kotyk, A., and Lev, A. A., *in* "Laboratory Techniques in Membrane Biophysics" (H. Passow and R. Stampfli, eds.) p. 69. Springer-Verlag, Berlin and New York, 1969.
61. Kuhbock, J., and Kummer, F., *Wien. Z. Inn. Med. Ihre Garenzgeb.* **53,** 513 (1972).
62. Lie, S. O., McKusick, V. A., and Neufeld, E. F., *Proc. Natl. Acad. Sci. U.S.A.* **69,** 2361 (1972).
63. Loewenstein, W. R., *Collect. Pap. Annu. Symp. Fundam. Cancer Res.* **28,** 239 (1975).
64. Maloney, P. C., Kashket, E. R., and Wilson, T. H., *Proc. Natl. Acad. Sci. U.S.A.* **71,** 3896 (1974).
65. Manfredi, F., *J. Lab. Clin. Med.* **61,** 1005 (1963).
66. Marton, L. J., Vaughn, J. G., Hawk, I. A., Levy, C. C., and Russell, D. H., *in* "Polyamines in Normal and Neoplastic Growth" (D. H. Russell, ed.), p. 367. Raven, New York, 1973.
67. Marvin, J. W., *Am. J. Bot.* **26,** 280 (1939).
68. Marvin, J. W., *Am. J. Bot.* **26,** 487 (1939).
69. Meyer, K. A., Kammerling, E. M., Amtman, L., and Koller, M., *Cancer Res.* **8,** 513 (1948).
70. Naeslund, J., and Swenson, K. E., *Acta Obstet. Gynecol. Scand.* **32,** 358 (1953).
71. Nakashima, M., Sousa, J. A., and Clapp, R. C., *Nature* (*London*) **235,** 16 (1972).
72. Needham, J., and Needham, D. M. *Proc. R. Soc. London, Ser. B* **98,** 259 (1925).
73. Needham, J., and Needham, D. M., *Proc. R. Soc. London, Ser. B* **99,** 173 (1926).
74. Obara, Y., Yoshida, H., Chai, L., Weinfeld, H., and Sandburg, A., *J. Cell Biol.* **58,** 608 (1973).
75. Paillard, M., *J. Physiol.* (*London*) **223,** 297 (1972).
76. Pandit, C. G., and Chambers, R., *J. Cell. Comp. Physiol.* **2,** 243 (1932).
77. Poole, D. T., Butler, T. C., and Waddell, W. J., *J. Natl. Cancer Inst.* **32,** 939 (1964).
78. Poole, D. T., Butler, T. C., and Williams, M. E., *J. Membr. Biol.* **5,** 261 (1971).
79. Racker, E., *Am. Sci.* **60,** 56 (1972).
80. Raven, J. A., and Smith, F. A., *in* "Ion Transport in Plants" (W. P. Anderson, ed.), p. 271. Academic Press, New York, 1973.
81. Reitnaur, P. G., *Z. Med. Labortech.* **13,** 5 (1972).
82. Ridgeway, E. G., Gilkey, J. C., and Jaffe, L. F., *J. Cell Biol.* **70,** 227a (1976).
83. Roos, A., *J. Physiol.* (*London*) **249,** 1 (1975).
84. Rubin, H., *J. Cell Biol.* **51,** 686 (1971).
85. Rubin, H., *in* Growth Control in Cell Cultures. *Ciba Found. Symp.*, p. 127 (1971).
86. Rubin, H., *J. Cell. Physiol.* **82,** 231 (1973).
87. Russell, D. H., *in* "Polyamines in Normal and Neoplastic Growth" (D. H. Russell, ed.), p. 2. Raven, New York, 1973.
88. Schuldiner, S., Rottenberg, H., and Avron, M., *Eur. J. Biochem.* **25,** 64 (1972).
89. Scott, G. D., and Kilgour, D. M., *J. Phys. D* **2,** 863 (1969).

89a. Shimada, T., and Ingalls, T. H., Arch. Environ. Health **30,** 196 (1975).
90. Siggaard-Andersen, O., "The Acid-Base Status of the Blood." Williams & Wilkins, Baltimore, Maryland, 1974.
91. Siskin, J. E., and Kinosita, R., *J. Biophys. Biochem. Cytol.* **9,** 509 (1961).
92. Smith, T. A., *Endeavour* **31,** 22 (1972).
92a. Taylor, A. C., *J. Cell Biol.* **15,** 201 (1962).
93. Thomas, R. C., *J. Physiol.* (*London*) **238,** 159 (1974).
94. von Ardenne, M., *Adv. Pharmacol. Chemother.* **10,** 339 (1972).
95. Waddell, W. J., and Butler, T. C., *J. Clin. Invest.* **38,** 720 (1959).
96. Walker, N. A., and Smith, F. A., *Plant Sci. Lett.* **4,** 125 (1975).
97. Warburg, O., *Science* **123,** 309 (1956).
98. West, I. C., *in* "Ion Transport in Plants" (W. P. Anderson, ed.), p. 237. Academic Press, New York, 1973.
99. Wiggins, P. M., *J. Theor. Biol.* **37,** 363 (1972).
100. Yamaha, G., *Cytologia* **6,** 523 (1935).

7

Cell Fusion and Regulation of DNA Synthesis

Potu N. Rao and Prasad S. Sunkara

I. INTRODUCTION

Nearly a quarter century after the delineation of the eukaryotic cell cycle into G_1, S, G_2, and M (14), relatively little is known about the molecular events that regulate the progression of cells from one phase to the next. An understanding of the cell cycle regulation would be invaluable in controlling or successfully treating malignant diseases in man. Apart from the two major events in the cell cycle, namely, DNA synthesis in S phase and chromosome condensation and division during mitosis, not much is known about the specific biosynthetic activities taking place during G_1 and G_2. The object of this article is to focus attention on the G_1 period, particularly in the light of evidence gathered by the use of techniques such as nuclear transplantation, cell fusion, and cytochalasin B-induced binucleation.

II. THE ROLE OF G_1 PERIOD IN CELL CYCLE

The pre-DNA-synthetic (G_1) period is usually considered as the time in the cell cycle when a cell is signaled to proceed for another round of DNA synthesis and subsequent mitosis (19). This is also the period when most of the preparations for the initiation of DNA synthesis are made by the cell. In cells having no G_1 period, such as *Amoeba proteus* and *Physarum polycephalum,* the metabolites necessary for the initiation of DNA synthesis are probably produced during the long G_2 phase (2, 3), suggesting that the molecular events necessary for the initiation of DNA synthesis need not necessarily have to follow mitosis, but in some cases they may precede it. Another important aspect of the G_1 period is the wide variability of its duration (28, 32) and its extreme sensitivity to unfavorable growth conditions (21), which usually leads to G_1 arrest of cells as a result of confluent culture conditions (17, 33, 35) or deprivation of certain essential nutrilites, such as leucine (5), isoleucine (16, 31) or tryptophan (1a).

In cultured mammalian cells, the duration of G_1 may range from 2 to 20 hours. There is also the example of a Chinese hamster cell line (V79) which is devoid of G_1 period (26). The fact that mammalian cells without a G_1 period can be found suggests that this period (G_1) is not very essential and can be eliminated without affecting the rest of the cell cycle. In those cells having a long G_1 period, the initiators of DNA synthesis were reported to be made toward the end of this phase (1, 10, 15, 18, 19, 29, 30, 34). However, inhibition of protein synthesis in mouse L cells for a fixed interval during early, middle, and late G_1 phase delayed the entry of these cells into S phase by a time equal to that of the duration of puromycin treatment (29, 30). These studies indicate that the three subdivisions of the G_1 phase were equally sensitive to the delaying effects of puromycin treatment and that uninterrupted protein synthesis is required for the completion of G_1 and the initiation of DNA synthesis.

Some of the recent studies have shown that the process of chromosome decondensation initiated at the telophase of mitosis continues throughout G_1 and even into S phase (13, 25, 27). In one of the experiments, a random population of HeLa cells prelabeled with [^{3}H]TdR was fused with a synchronized mitotic population to induce premature chromosome condensation in the former (25). The morphology of the prematurely condensed chromosomes (PCC) of the G_1, S, and G_2 phase cells was studied. The mor-

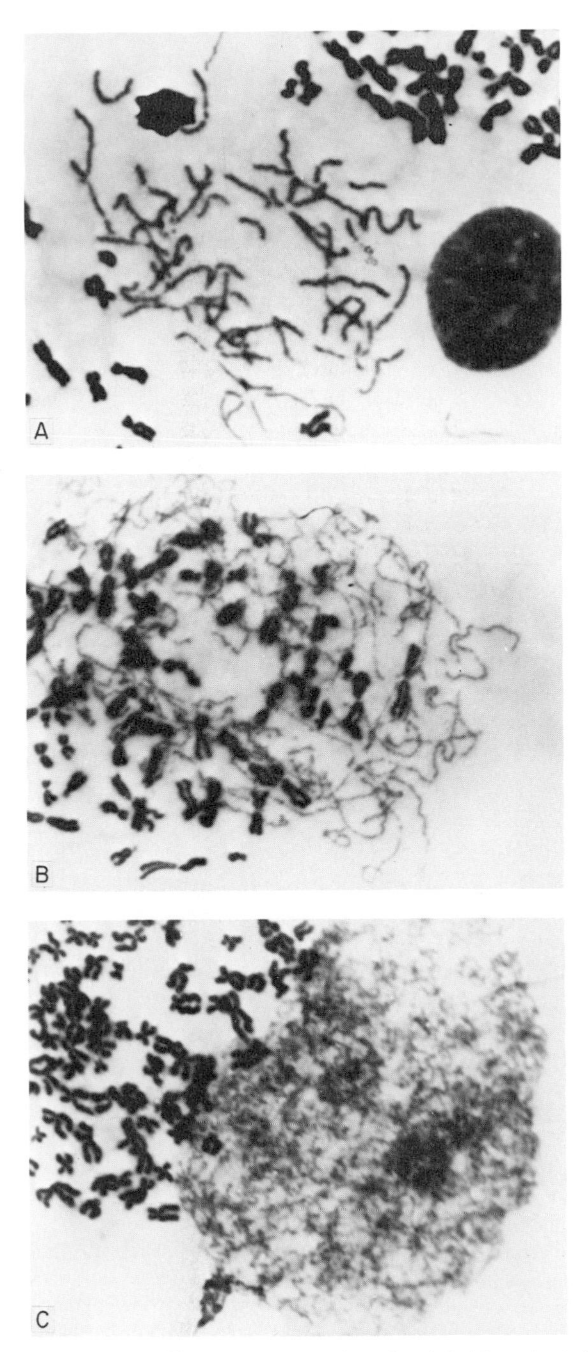

FIG. 1. G_1-PCC of HeLa cells. The highly condensed and darkly stained chromosomes are of mitotic cells. Note the variation in the morphology of the PCC. (A) Condensed PCC. (B) Less condensed PCC. (C) Extended PCC. [From Rao *et al.* (25).]

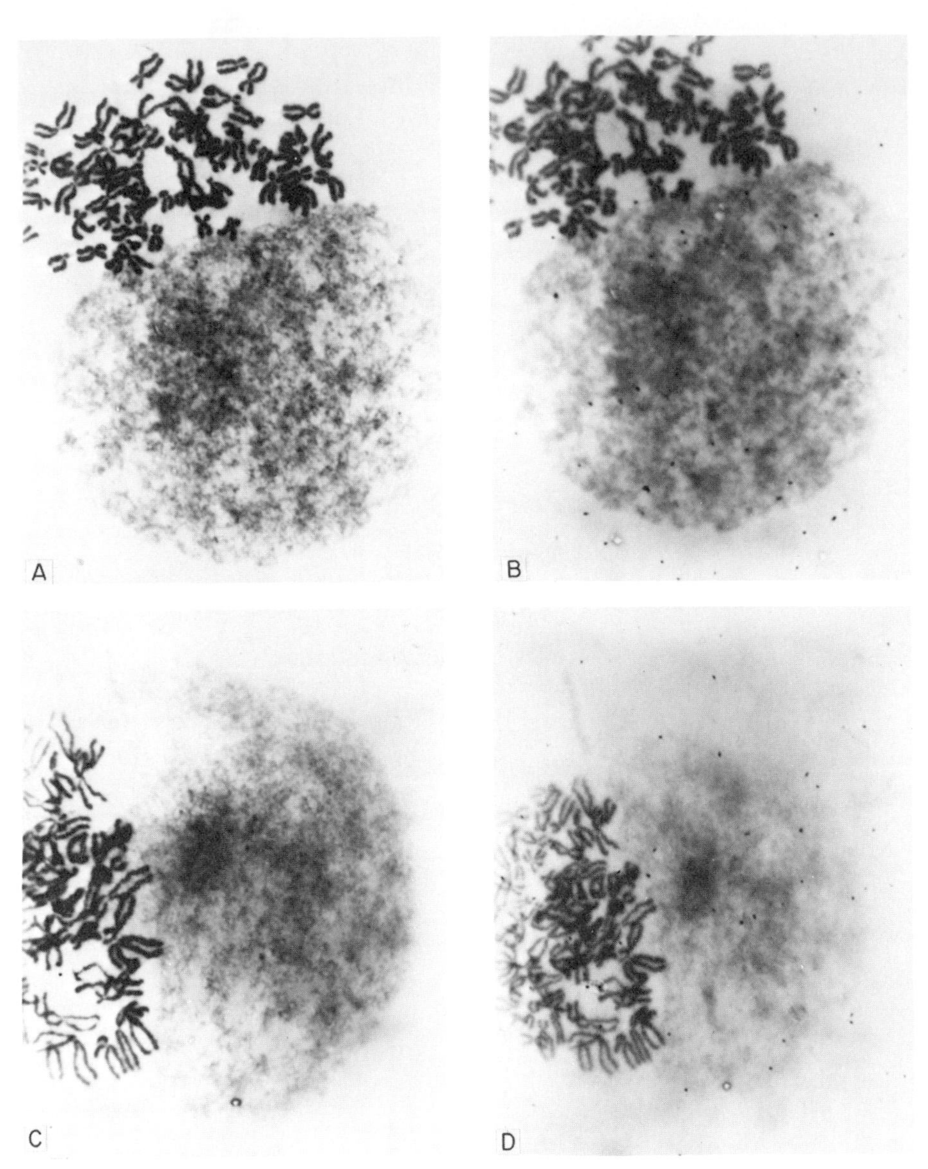

FIG. 2. G_1-PCC of HeLa cells. (A) PCC are greatly extended. (C) Very diffused G_1-PCC. (B) and (D) are autoradiographs of cells shown in (A) and (C), respectively. The grain densities on the PCC are no greater than the background indicating that these cells did not incorporate [^{3}H]TdR during a 15-minute pulse labeling immediately before fusion with mitotic cells. The gradual decondensation of G_1-PCC shown in this and the previous figure (Fig. 1) indicate the progressive changes in the conformational pattern of chromatin during the G_1 period. Very early G_1 cells yield condensed PCC, whereas late G_1 cells produce PCC with very diffused morphology. [From Rao *et al.* (25).]

phology of the G_1-PCC was found to be most variable, ranging from the most condensed to the most diffused state (Figs. 1 and 2). The PCC of early G_1 cells were well condensed and distinct in appearance with a single chromatid (Fig. 1A, B). The PCC of cells in the middle G_1 period tended to become more diffused, but still the single chromatid threads were discernible (Fig. 1C). However, in the late G_1 cells, the PCC became extremely diffused and indistinguishable from the early S-PCC except by autoradiography (Fig. 2). These data suggest that the process of progressive decondensation of chromatin is one of the characteristic events of G_1 phase that precedes the onset of DNA synthesis.

III. INDUCERS OF DNA SYNTHESIS IN S PHASE CELLS

Synchrony in the initiation of DNA synthesis in the nuclei of the nuturally occuring multinucleate organisms, such as the slime mold *Physarum*, amoeba *Pelomyxa*, and the ciliate *Urostyla*, is more common than asynchrony. In *Physarum*, when plasmodia of different states are brought in contact with each other, the nuclear synchrony with regard to DNA synthesis and mitosis is rapidly achieved (11, 12). Even in the artificially created multinucleate systems, for example, the caffeine-induced multinucleate onion root cells (7) and the cytochalasin B-induced multinucleate mammalian cells (6) all the nuclei within a cell entered S phase synchronously. The studies of Rao and Johnson (22) and Graves (9) have clearly demonstrated that DNA synthesis can be readily induced in G_1 nuclei following Sendai virus-mediated fusion between G_1 and S phase cells of the same or different species. The fusion of G_2 phase cells with either G_1 or S phase cells did not inhibit either the initiation or the continuation of DNA synthesis, indicating that there are no inhibitors of DNA synthesis in the G_2 cells.

These studies also indicate that DNA synthesis is regulated by a positive mode of control (22). These cell fusion studies confirm the earlier observations of de Terra (4), who reported the initiation of DNA synthesis in the G_1 nuclei of the ciliate *Stentor* after transfer into S phase cells. Similarly, Graham *et al.* (8) observed that DNA synthesis was initiated in the nuclei isolated from nondividing cells, such as adult liver, brain, and blood cells within 90 minutes after their transplantation into the unfertilized eggs of *Xenopus*. From all these observations, it becomes obvious that the inducers of DNA synthesis are present in the cytoplasm of the S phase cells.

IV. REGULATION OF DNA SYNTHESIS IN CYTOCHALASIN B-INDUCED BINUCLEATE CELLS

Do the binucleate cells have a shorter G_1 period, as compared to the mononucleate cells? The cell fusion studies of Rao and Johnson (22) and Graves (9) showed that binucleate cells produced by fusion between two G_1 cells entered S phase at about the same rate as mononucleate cells. However, the recent studies by Fournier and Pardee (6) on Syrian hamster BHK21/C13 cells indicated that binucleate cells, produced by treating mitotic cells with cytochalasin B for 90 minutes, completed G_1 period significantly faster than their mononucleate counterparts. To explain these results, they proposed a model, according to which there is a nuclear cooperation for the initiation of DNA synthesis, which is non-concentration-dependent. They assumed that, among the multinucleate cells, each nucleus makes a critical substance during early G_1 which is utilized during late G_1, and the total amount, not the concentration, of this substance determines the duration of G_1. The greater the amount of this substance available, the shorter is the G_1 period. Because binucleate cells make twice the amount of this substance than the mononucleate cells they have shorter G_1 period.

To test the above model, we repeated these studies with HeLa cells (23). A spinner culture of HeLa cells was first partially synchronized in S phase with an excess TdR (2.5 m*M*) block for 20 hours. After the release of the TdR block, the cells were plated and incubated at 37°C in a stainless steel chamber filled with N_2O at 80 psi (20). At the end of the N_2O block, the floating and loosely attached mitotic cells were harvested by gentle pipetting. They had a mitotic index of 98%. Mitotic cells thus collected were exposed to cytochalasin B (2 μg/ml) for 90 minutes. In order to study the effect of timing of the CB treatment on cell cycle kinetics, CB treatments were given either immediately or 90 minutes after the reversal of the N_2O block. Cells not exposed to CB served as control. The increase in labeling index of the mono-, bi-, tri-, and tetranucleate cells is plotted as a function of time (Fig. 3). These results can be summarized as follows.

In the control (not treated with CB), the binucleate cells, which constituted about 13% of the population, had a slightly (0.5 hours) shorter G_1 period than the mononucleate cells (Table I and Fig. 3A). However, the duration of G_1 period in the binucleate cells, produced by a CB treatment given immediately after the reversal of the N_2O block, was 1.6 hours shorter than that of mononucleate cells. Trinucleate cells reached a labeling index of 50% only slightly (about 0.2 hours) ahead of the binucleate cells, and a similar difference was seen between tetra- and trinucleate cells (Fig. 3B). The bi- and trinucleate cells, produced by a CB treatment given at 90 minutes after the

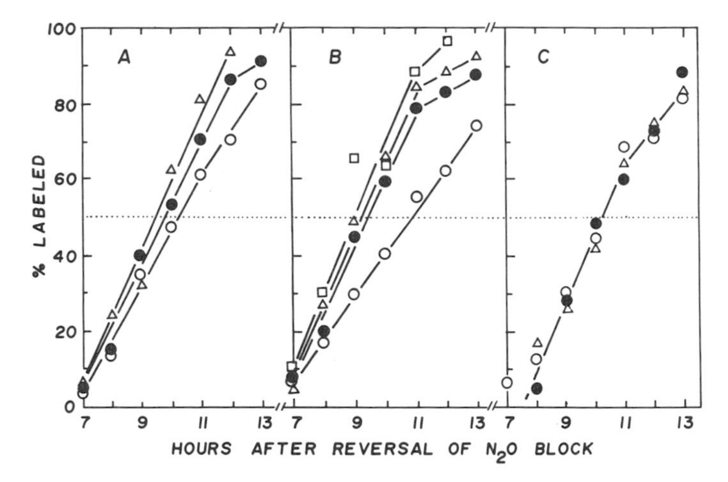

FIG. 3. The rate of initiation of DNA synthesis as a function of time after the reversal of the N_2O block. The 50% labeling index is indicated by the dotted line. ○, mono-; ●, bi-; △, tri-; and □, tetranucleate cells. (A) Control; CB treatment. (B) 0–90 minutes and (C) 90–180 minutes after the reversal of the mitotic block. [From Rao and Smith (23).]

reversal of the N_2O block, entered S phase at the same rate as mononucleate cells (Fig. 3C).

In other words, the time of application of CB treatment after the reversal of the mitotic block has a profound effect on the rate of entry of the mono- and binucleate cells into S phase. This was found to be due to the degree of asynchrony in the division of mitotic cells (23). Even in a highly

TABLE I

Rate of Initiation of DNA Synthesis in HeLa Cells after Cytochalasin-B Treatment[a]

			Treatments		
			A	B	C
Duration (in minutes) of CB treatment after the reversal of N_2O block			0 (control)	0–90 minutes	90–180 minutes
Frequency of	(a)	Mononucleate cells	83.0	26.6	74.4
	(b)	Binucleate cells	13.5	43.6	18.4
	(c)	Multinucleate cells	3.5	29.8	7.2
Time (hr) after reversal of N_2O block, required to achieve a labeling index of 50% in	(a)	Mononucleate cells	10.2	10.9	10.5
	(b)	Binucleate cells	9.7	9.3	10.3
	(c)	Difference (a − b)	0.5	1.6	0.2

[a] Adapted from Rao and Smith (23).

synchronized mitotic population, all the cells do not complete division at the same time. Naturally, the cells that completed mitosis earlier would enter S phase earlier than the late dividers. Addition of CB during the first 90 minutes after the reversal of the mitotic block would induce binucleation among the early dividing fraction, and, hence, the binucleate cells resulting from that treatment would enter S phase earlier than the mononucleates. Conversely, the application of CB at 90 minutes after reversal of the mitotic block would induce binucleation in the late dividers, and these binucleate cells would enter S phase at about the same rate as the mononucleate cells (Fig. 3C). From these studies, it appears that the differences in the rates of entry into S phase of the mono- and binucleate cells was largely due to an experimental artifact. These studies also reveal that binucleate cells have a slightly shorter G_1 period than mononucleate cells, but the difference is usually less than 0.5 hour. Thus, the results of Fournier and Pardee (6) and Rao and Smith (23) can be explained without invoking the model for non-concentration-dependent cooperative initiation of DNA synthesis. But supporting evidence for this model comes from another study of ours (24), which is discussed in the following section.

V. G_1 PERIOD AND THE INDUCERS OF DNA SYNTHESIS

It is generally understood that the inducers of DNA synthesis are made towards the end of G_1 period (19). But so far we have no experimental evidence to suggest whether the inducers accumulate gradually during the G_1 or become suddenly available at the G_1-S transition. Our recent studies have provided a definite answer to this question (24).

In these studies, we have obtained highly synchronized populations of early, middle, and late G_1 HeLa cells by harvesting them at 3, 5, and 7 hours, respectively, after the reversal of a mitotic block by N_2O at high pressure. For this study, six different fusions were made. They were between early G_1 and late G_1 [(A)EG_1/LG_1* and (B)EG_1*/LG_1]; early G_1 and mid-G_1 [(C)EG_1/MG_1* and (D)EG_1*/MG_1] and mid-G_1 and late G_1 [(E)MG_1/LG_1* and (F)MG_1*/LG_1]. The asterisk indicates the population which was prelabeled with [^{3}H]TdR. To study the cell cycle progression of both the parents and compare them with that of hybrids, it was essential to prelabel one parent at a time. Immediately after fusion, the cells were plated in dishes with culture medium containing [^{3}H]TdR and incubated at 37°C. Samples were taken at regular intervals by trypsinizing one dish for each fusion. Cells deposited directly on slides by the use of a cytocentrifuge were fixed, processed for autoradiography, stained with Giemsa, and scored for the percent of increase in the labeling index of the various types of fused cells (for details, see Rao *et al.*, 24).

Before we present the results, it is important to know the various classes of cells present in the fusion mixture. In each fusion the mono-, bi-, and trinucleate cells were scored. Cells with four or more nuclei were not considered because of their low frequency. The following are the different classes of cells scored. They are: (1) mononucleate cells: labeled (L) and unlabeled (U); (2) binucleate cells: binucleates with nuclei of the same age or NSA binucleates (LL and UU) and binucleates with nuclei of different ages or NDA-binucleates (LU); (3) trinucleate cells (LLL, LLU, LUU, and UUU).

A. Fusions between Early G_1 and Late G_1

The kinetics of initiation of DNA synthesis in the mono-, bi-, and trinucleate cells resulting from fusions between early G_1 and late G_1 are shown in Fig. 4. In fusion A, the late G_1 cells were prelabeled, whereas, in fusion B, early G_1 cells were prelabeled. The rate of entry of the NSA-binucleate cells (EG_1/EG_1) into S phase is not significantly different from that of the mononucleate cells (EG_1). However, the early G_1 nuclei located in the NDA-binucleate cells (EG_1/LG_1) entered S phase at the same time as the LG_1 nucleus (Fig. 4A). The NSA-binucleate cells of the late G_1 parent (LG_1/LG_1) traversed G_1 phase significantly faster than the mononucleate (LG_1) cells (Fig. 4B). The NSA-trinucleate cells (3 LG_1) entered S phase even faster than the NSA-binucleates (LG_1/LG_1). The NDA-binucleate cells (EG_1/LG_1) exhibited a G_1 period which was a little shorter than that of mononucleate LG_1 cells (Fig. 4B).

B. Fusions between Early G_1 and Mid-G_1

In fusions involving early and mid-G_1 populations, mid-G_1 cells were prelabeled for fusion C and early G_1 cells were prelabeled for fusion D. The kinetics of the initiation of DNA synthesis are shown in Fig. 5. The NDA-binucleate cells (EG_1/MG_1) reached a labeling index of 50% at 9.5 hours, as compared to the 12.75 hours for the mono- and the NSA-binucleate (EG_1/EG_1) cells (Fig. 5A). The NSA-binucleates of the mid-G_1 parent (MG_1/MG_1) reached the 50% LI mark about 1 hour earlier than their mononucleate (MG_1) counterparts (Fig. 5B). The NSA-trinucleates of the mid-G_1 parent (3 MG_1) completed G_1 period even faster than the NSA-binucleate (MG_1/MG_1) cells.

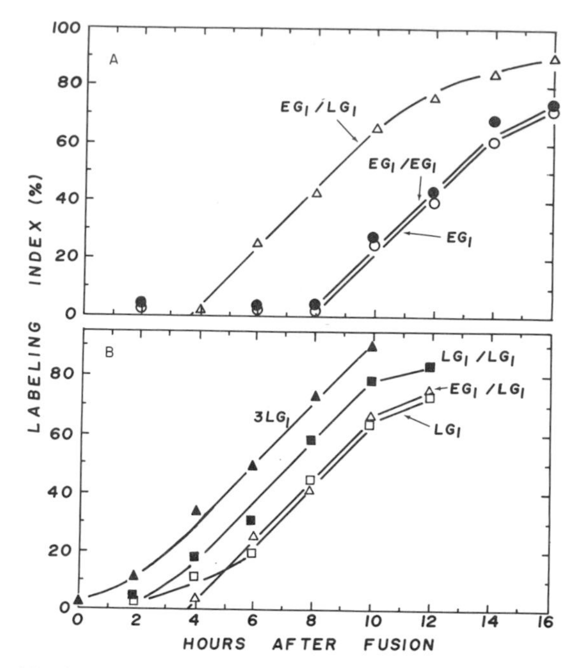

FIG. 4. The kinetics of the initiation of DNA synthesis in fustions between early G_1 and late G_1 populations. (A) These data are from fusion A (EG_1/LG_1*). The G_1 traverse, measured as increase in labeling index as a function of time, of the early G_1 nuclei residing in mononucleate (EG_1), ○; NSA-binucleate (EG_1/EG_1), ●; and NDA-binucleate (EG_1/LG_1), △, cells are shown in this panel. (B) These data are from fusion B (EG_1*/LG_1). The rates of increase in the labeling index of the late G_1 nuclei residing in mono-, bi-, and tri-nucleate cells are presented in this graph: □, mononucleate (LG_1); ■, NSA-binucleate (LG_1/LG_1); △, NDA-binucleate (EG_1/LG_1); and ▲, NSA-trinucleate ($3LG_1$). [From Rao *et al.* (24).]

C. Fusions between Mid-G_1 and Late G_1

The NDA-binucleate cells (MG_1/LG_1) in fusion E entered S phase earlier than the mono- and NSA-binucleate (MG_1/MG_1) cells (Fig. 6A). This is due to the premature induction of DNA synthesis in MG_1 nucleus, which was influenced by the entry of LG_1 component into S phase. In fusion F, involving labeled MG_1 and unlabeled LG_1, the NDA-binucleates (MG_1/LG_1) reached the 50% LI mark significantly earlier than the mononucleate LG_1 cells but were only slightly behind the NSA-binucleate (LG_1/LG_1) cells (Fig. 6B). The times required to reach the 50% LI mark for the different types of cells in all these fusions (A–F) are shown in Table II, which is in essence a summary of Figs. 4, 5, and 6.

D. The Time of Synthesis of the Inducers of DNA Synthesis

The above studies clearly show that, when two mid or late G_1 cells are fused together, they traverse G_1 period faster than the unfused mononucleate cells. This suggests that G_1 cells, upon fusion, cooperate with each other by pooling their resources for the purpose of the initiation of DNA synthesis. What are these resources? They are probably proteins that can initiate DNA synthesis. According to Table II, the fusion of one late G_1 cell with another would reduce the duration of G_1 period by about 1.5 hours. On the other hand, the addition of one mid-G_1 cell to either late G_1 or mid-G_1 cell would shorten the G_1 by 1 hour only. When an early G_1 cell is fused with MG_1 or LG_1, the reduction in the length of G_1 is about 0.25 hour. So it appears that the farther a cell had advanced in G_1, the greater would be its contribution for the initiation of DNA synthesis. Hence, the levels of the

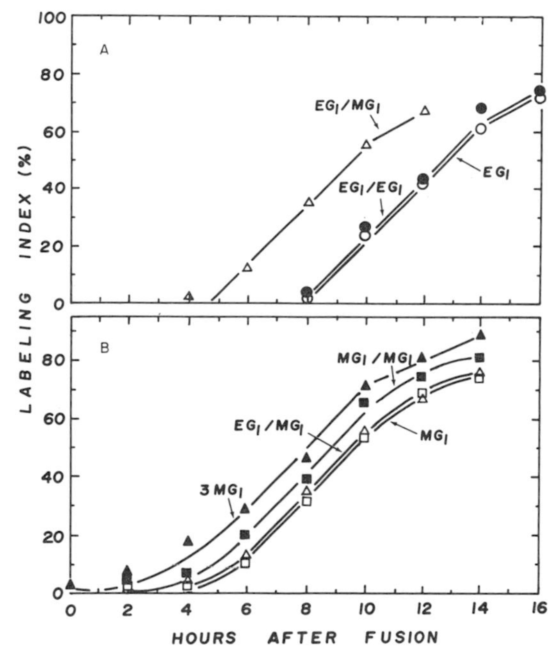

FIG. 5. The kinetics of the initiation of DNA synthesis in fusions between early G_1 and mid-G_1 cells. (A) These data are from fusion C (EG_1/MG_1*). The rate of entry of the unlabeled early G_1 nuclei into S phase was plotted as a function of time. ○, mononucleates (EG_1); ●, NSA-binucleate (EG_1/EG_1); and △, NDA-binucleates (EG_1/MG_1). (B) These data are from fusion D (EG_1*$/MG_1$). The increase in the labeling index of the mid-G_1 nuclei was plotted as a function of time. □, mononucleates (MG_1); ■, NSA-binucleates (MG_1/MG_1); △, NDA-binucleates (EG_1/MG_1); and ▲, NSA-trinucleates ($3MG_1$). [From Rao *et al.* (24).]

inducers of DNA synthesis in late G_1 cells are significantly higher than those in middle or early G_1 cells.

The fact that the reduction of the length of G_1 period of a late G_1 cell is proportional to the number and age of G_1 cells added to it by fusion indicates that the inducers of DNA synthesis operate in a non-concentration-dependent manner. In other words, the total number of inducer molecules present within a cell are more important for the initiation of DNA synthesis, rather than their concentration in the cytoplasm or the nucleus. When more cells are fused together, the total number of the inducer molecules increases without any significant changes in their concentration. When a critical number is reached, DNA synthesis is initiated in one or more nuclei at the same time. The number of nuclei present within a cell does not appear to have an affect on the rate of initiation.

On the basis of these observations, a model has been proposed for the availability of the inducers of DNA synthesis during the HeLa cell cycle

TABLE II

A Comparison of the Rates of Entry of Mononucleate, NSA- and NDA-Binucleate Cells into S Phase Measured (As Hours) after Fusion to Reach a Labeling Index of 50%[a]

	Cell Type	50% LI at (hr)	Relative contribution (hour of advancement) for each additional cell in		
			LG_1	MG_1	EG_1
a	LG_1	8.75			
b	$LG_1 + LG_1$	7.25			
	(a–b)	1.5	1.5		
c	3 LG_1	6.0			
	(a–c)	2.75	1.4		
d	MG_1	9.75			
e	$MG_1 + MG_1$	8.75			
	(d–e)	1.0		1.0	
f	3 MG_1	8.0			
	(d–f)	1.75		0.9	
g	$LG_1 + MG_1$	7.75			
	(a–g)	1.0		1.0	
h	EG_1	12.75			
i	$EG_1 + EG_1$	12.7			
	(h–i)	0.05			0.05
j	$LG_1 + EG_1$	8.5			
	(a–j)	0.25			0.25
k	$MG_1 + EG_1$	9.5			
	(d–k)	0.25			0.25

[a] From Rao *et al.* (24).

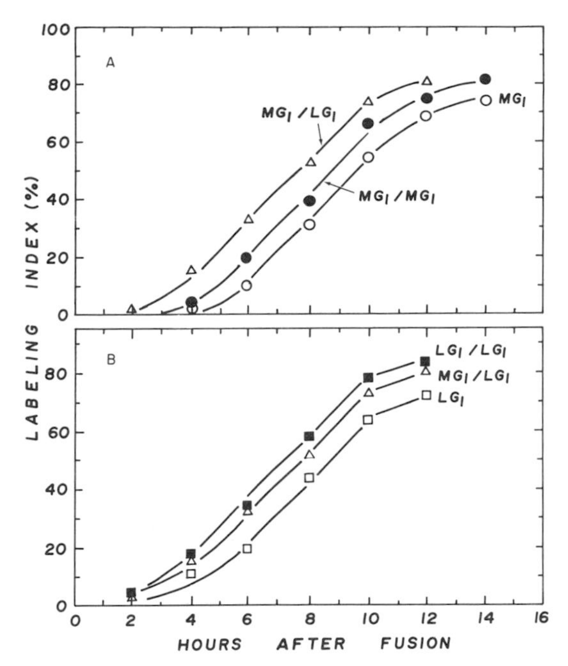

FIG. 6. The kinetics of the initiation of DNA synthesis in fusions between mid-G_1 and late G_1 populations. (A) These data are from fusion E (MG_1/LG_1*). The progress of mid-G_1 nuclei into S phase was plotted. ○, mononucleates (MG_1); ●, NSA-binucleates (MG_1/MG_1); △, NDA-binucleates (MG_1/LG_1). (B) These data are from fusion F (MG_1*/LG_1). The G_1 traverse of late G_1 nuclei is shown in this panel. □, mononucleate (LG_1); ■, NSA-binucleates (LG_1/LG_1); and △, NDA-binucleates (MG_1/LG_1). [From Rao *et al.* (24).]

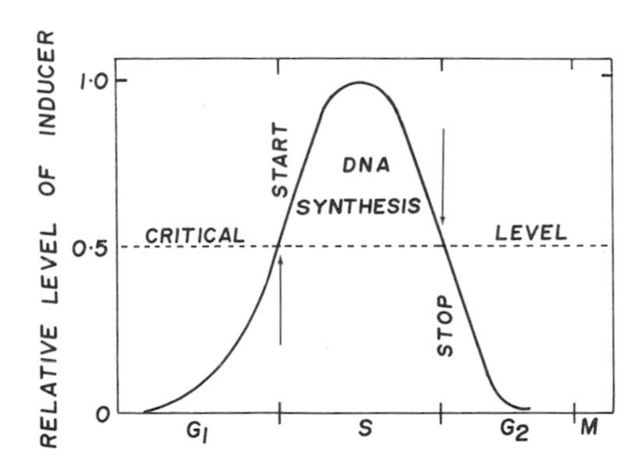

FIG. 7. A model to explain the appearance of the inducer(s) of DNA synthesis during the HeLa cell cycle. The relative levels (arbitiary units) of the inducer is shown to vary according to the position of a cell in the cell cycle. G_1, pre-DNA synthetic period; S, DNA synthetic period; G_2, the post-DNA synthetic period; M, mitosis. The duration of these various phases are not drawn to scale. [From Rao *et al.* (24).]

(24). According to this model, the inducers accumulate gradually throughout the G_1 period, reaching a critical level at the G_1-S transition when the DNA replication is initiated (Fig. 7). The level of the inducers probably reaches a peak during the early or middle S phase and then gradually decline below the critical level at the S–G_2 boundary when DNA synthesis ceases.

VI. CONCLUSIONS

1. The inducers of DNA synthesis accumulate gradually throughout the G_1 period, reaches a maximum level in S phase, and then gradually decline in G_2.
2. The initiation or the continuation of DNA synthesis can occur only when the inducers are present above the critical level, which coincides with the beginning and the end of S phase.
3. The inducers of DNA synthesis operate in a non-concentration-dependent manner. The absolute number of the inducer molecules is critical for the initiation DNA synthesis, but not their concentration within a cell.

ACKNOWLEDGMENTS

This study was supported in part by Grants CA-16480, CA-11520, CA-14528, and CA-19856 from the National Cancer Institute, DHEW; Grant GM-23252 from the Institute of General Medical Sciences, DHEW; and Grant VC-163 from the American Cancer Society.

REFERENCES

1. Baserga, R., *Cell Tissue Kinet.* **1,** 167 (1968).
1a. Brunner, M., *Cancer Res.* **33,** 29 (1973).
2. Cummins, J. E., Blomquist, J. C., and Rusch, H. P., *Science* **154,** 1343 (1966).
3. Cummins, J. E., Brewer, E. N., and Rusch, H. P., *J. Cell Biol.* **25,** 337 (1965).
4. de Terra, N., *Proc. Natl. Acad. Sci. U.S.A.* **57,** 607 (1967).
5. Everhart, L. P., and Prescott, D. M., *Exp. Cell Res.* **75,** 170 (1972).
6. Fournier, R. E., and Pardee, A. B., *Proc. Natl. Acad. Sci. U.S.A.* **72,** 869 (1975).
7. González-Fernandéz, A., Giménez-Martin, G., Diez, J. L., de la Torre, C., and López-Sáez, F. J., *Chromosoma* **36,** 100 (1971).
8. Graham, C. F., Arms, K., and Gurdon, J. B., *Dev. Biol.* **14,** 349 (1966).
9. Graves, J. A. M., *Exp. Cell Res.* **72,** 393 (1972).
10. Gurdon, J. B., and Woodland, H. R., *Biol. Rev.* **43,** 233 (1968).
11. Guttes, E., Devi, V. R., and Guttes, S., *Experientia* **25,** 615 (1969).
12. Guttes, E., Guttes, S., and Rusch, H. P., *Fed. Proc., Fed. Am. Soc. Exp. Biol.* **18,** 479 (1959).

13. Hittelman, W. N., and Rao, P. N., *Exp. Cell Res.* **100,** 219 (1976).
14. Howard, A., and Pelc, S. R., *Heredity* **6,** 261 (1953).
15. Lark, K. G., *Mol. Genet.* **1,** p. 153 (1963).
16. Ley, K. D., and Tobey, R. A. *J. Cell Biol.* **47,** 453 (1970).
17. Nilausen, K., and Green, H., *Exp. Cell Res.* **40,** 166 (1965).
18. Prescott, D. M., *Cancer Res.* **28,** 1815 (1968).
19. Prescott, D. M., *Adv. Genet.* **18,** 99 (1976).
20. Rao, P. N., *Science* **160,** 774 (1968).
21. Rao, P. N., and Engelberg, J., *in* "Cell Synchrony: Studies in Biosynthetic Regulation" (I. L. Cameron and G. M. Padilla, eds.), p. 332. Academic Press, New York, 1966.
22. Rao, P. N., and Johnson, R. T., *Nature* (*London*) **225,** 159 (1970).
23. Rao, P. N., and Smith, M. L., *Exp. Cell Res.* **103,** 213 (1976).
24. Rao, P. N., Sunkara, P. S., and Wilson, B. A., *Proc. Natl. Acad. Sci. U.S.A.* **74,** 2869 (1977).
25. Rao, P. N., Wilson, B. A., and Puck, T. T., *J. Cell. Physiol.* **91,** 131 (1977).
26. Robbins, E., and Scharff, M.D., *J. Cell Biol.* **34,** 684 (1967).
27. Schor, S. L., Johnson, R. T., and Waldren, C. A., *J. Cell Sci.* **17,** 539 (1975).
28. Sisken, J. E., and Kinosita, R., *J. Biophys. Biochem. Cytol.* **9,** 509 (1961).
29. Terasima, T., and Yasukawa, M., *Exp. Cell Res.* **44,** 669 (1966).
30. Terasima, T., Fujiwara, Y., Tanaka, S., and Yasukawa, M., *in* "Cancer Cells in Culture" (H. Katsuta, ed.), p. 73. Univ. Park Press, Baltimore, Maryland, 1968.
31. Tobey, R. A., *Methods Cell Biol.* **6,** 67 (1973).
32. Tobey, R. A., Anderson, E. C., and Petersen, D. F., *J. Cell Biol.* **35,** 53 (1967).
33. Todaro, G. J., and Green, H., *J. Cell Biol.* **17,** 299 (1963).
34. Webster, P. L., and Van't Hoff, J., *Exp. Cell Res.* **55,** 88 (1969).
35. Yoshikura, H., and Hirokawa, Y., *Exp. Cell Res.* **52,** 439 (1968).

8

Regulation of Gene Expression in the Cell Cycle of *Physarum*

Helmut W. Sauer

I. INTRODUCTION

An understanding of the regulation of gene expression is probably the most important unsolved problem in cell biology. Although our knowledge of the structural organization of chromatin, multiple RNA polymerases, and RNA transcription products in eukaryotic cells is progressing rapidly, a control mechanism of transcription cannot be convincingly formulated. Even the central hypothesis of differential transcription may have to be substituted for a selective stabilization of some RNA species over others. Therefore, it comes as no surprise when D. M. Prescott states, in a recent review of the cell cycle and the control of cellular reproduction, that virtually nothing is known about the significance of RNA transcription in the cell cycle and calls for analyses of synchronous systems (1).

In the following sections, some results on gene expression obtained with *Physarum*, a synchronous system, will be presented, and some speculations on possible transcription control mechanisms in this organism will be made.

II. THE CELL CYCLE OF *PHYSARUM*

Myxomycetes are characterized by a plasmodial organization during the vegetative phase of their well-defined life cycles (2, 3). Significantly, up to 10^8 nuclei undergo mitosis in regular intervals (8–10 hours at 26°C) in the common cytoplasm of a macroplasmodium of *Physarum polycephalum*. The natural mitotic synchrony, together with the development of liquid culture media, have made this organism an attractive model system for studies of the structure and biochemistry of this giant "cell," as well as for experiments designed to learn something about the regulation of a mitotic cycle.

Earlier work on *Physarum* has been summarized by Rusch (4); some aspects of differentiation have been discussed by Sauer (5); more recent data on the cell cycle (6), nuclei (7), nuclear protein (8, 9), and RNA (10) have also been reviewed.

In three aspects, the cell cycle of *Physarum* is atypical: (a) mitosis is intranuclear; (b) cytokinesis does not follow nuclear division; and (c) S phase starts in telophase, leaving no room for a G_1 phase. One may rationalize these facts in that nuclear division in *Physarum* displays regulation of mitosis in its purest form, undisturbed by (a) drastic cell rearrangements connected with nuclear membrane breakdown, (b) physiological requirements for cell division, and (c) the complex regulatory mechanism of however, like (a) the problems with assigning a certain G_2 phase event with either mitosis or S phase, and (b) the question—in view of other examples of cell types having no G_1 phase (see Prescott), whether G_1 and G_2 phases are at all distinct and essential parameters of the cell cycle or, more specifically, the DNA division cycle (11). Therefore, it is a challenge for "physarologists" to establish whether (a) mitosis and S phase can be uncoupled, and (b) whether replication can be obtained in G_2 phase without prior mitosis.

The most conspicuous markers of the cell cycle of *Physarum* are morphological changes (of nucleolus, chromatin, and spindle structures) around the time of mitosis, and the restriction of chromosomal DNA replication to S phase. In addition, there are a number of events (summarized in the previous reviews) that may be specific for the cell cycle, such as (a) enzyme activity peaks (thymidine kinase, histone kinase, NAD phosphorylase); steps (RNase, DNA polymerase), and even a valley (polyADPribose polymerase), (b) accumulation of actin in the nucleolus, (c) an increase in the phosphate content of H1 histone, (d) sensitivity to temperature shocks, and (e) the requirement for protein synthesis. Mutants will eventually help to prove which of these observations define essential transition points (12). So far, we have learned that mutants without thymidine kinase grow normally (Haugli, personal communication).

III. GENE EXPRESSION IN THE CELL CYCLE

Although there is no evidence for the requirement of a specific transcriptional event in any cell cycle, mitotic delay observed in *Physarum* after the addition of actinomycin C up to 30 minutes before mitosis may indicate some role for RNA molecules (13). However, it has not been possible to discriminate between RNA classes by *in vivo* application of cordycepin and amatoxin (14). Therefore, a better understanding of the necessity of transcription will come from studies of transcription products. In the following sections, I shall describe the transcription machinery (DNA, RNA polymerases, and RNA molecules) and comment on RNA synthesis *in vivo* and *in vitro* of various cell cycle stages.

A. The Transcription Machinery of *Physarum*

1. Templates

Physarum polycephalum contains three classes of DNA which can be distinguished according to location, buoyant density, and molecular size. Mitochondrial DNA (1,688 gm $\times$ cm^{-3}, 40 $\times$ 10^{6} daltons) accounts for approximately 8% of total DNA, replicates throughout the cell cycle (for review, see 4), and contains open DNA circles of about 19 μm (15).

Nucleoli contain 2% of the total DNA, which makes up the heavy satellite in CsCl gradients (1,712 gm $\times$ cm^{-3}). Replication of rDNA occurs throughout the cell cycle, with the possible exception of early S phase. The exact structure of rDNA is not known, and linear molecules of 40 $\times$ 10^{6} daltons (each possibly a palindrome with two terminal cistrons) have been described (16), as well as circles (of 8 $\times$ 10^{6} daltons, and multiples thereof) and lariats (17), reminiscent of the situations in *Tetrahymena* (18) and oocytes (19).

However, it seems clear that most ribosomal genes are not integrated in the more complex chromosomal DNA, and they behave in some respects like episomes. Together with the ease of nucleolar isolation and preliminary results of *in vitro* rDNA replication and transcription, the study of *Physarum* may provide yet another interesting system—naturally cloned genes, complete with regulative base sequences and a high concentration of putative regulatory proteins due to the multiple copies of ribosomal genes (150–250 per nucleus).

Chromosomal DNA (1,702 gm $\times$ cm^{-3}) approximately 1 pg per diploid nucleus of *Physarum,* is contained in 70–80 chromosomes (7). This DNA is about 100 times more complex than that of *Escherichia coli* (20), which puts *Physarum* in the vicinity of lower chordates, rather than solitary or social amoebae. Chromosomal DNA replication, as measured by thymidine incorporation is restricted to S phase (21), occurs in small pieces, which contain short stretches of RNA in covalent linkage (22) and is ligated, possibly stepwise, to replicon size (35 S) (23). Maturation of very large DNA, probably one molecule per chromosome, occurs during G_2 phase (F. Haugli, personal communication).

Discontinuous replication is also suggested from preliminary DNA fiber autoradiographs. We have seen short stretches (2,4,8 μm, 5-minute pulse) and longer ones (25, 50, 75, 500 μm, 45-minute pulse). Putative replicating DNA molecules detected in formamide spreads contain variable "bubbles" ranging from 0.5 to 25 μm. Several examples show a small bubble adjacent to a large one on the same molecule. More work is necessary before we can distinguish whether equal size replicons replicate asynchronously or replicons of different size replicate synchronously. Nevertheless, it has been elegantly shown, several times (for review, see 6) that chromosomal DNA replicates sequentially. It is likely that up to 10 banks of replicons synthesize DNA successively, possibly controlled by concomitant protein synthesis (6, 24). Besides a temporal sequence, it seems that early replicating DNA differs from late DNA in that is more dense (25) and contains less repeated base sequences (20, 26). This might indicate that euchromatin of *Physarum* replicates before heterochromatin, as is known in other organisms (for review, see 1).

Physarum nuclei contain histones that resemble other histones (27), and there is recent evidence that nucleosomes can be generated by nuclease treatment, even from metaphase nuclei (28). A preliminary analysis of chromatin has revealed a high protein/DNA ratio (29) of about 4. This may indicate an unusually high content of nuclear proteins, many of which may be due to nucleolar contamination.

It may be of interest for discussions below that (1) replicating DNA is more sensitive to DNase I than interphase DNA, and that (2) the migration of isolated nuclei in an electric field is determined by DNA and is especially high in early S phase (our unpublished observations).

2. RNA Polymerases

There are four different RNA polymerases in *Physarum,* as in other cells. Two enzymes have been purified extensively and can be identified as form A in the nucleolus and form B in the nucleoplasm (30–32). The third enzyme although unstable, has been characterized by its chromatographic behavior, sensitivity to high doses of amatoxin, and resistance to rifampicin as a C-type enzyme (33). The fourth enzyme is rifampicin-sensitive, has been detected in mitochondrial extracts, and may be a mitochondrial RNA polymerase (34).

a. Solubilized RNA Polymerases. The well-analyzed RNA polymerases (A and B) have a slow turnover. Therefore, their rates of synthesis probably cannot regulate transcription during the cell cycle (35). However, there is a drastic and selective reduction of the RNA polymerase A level during starvation (35), and some evidence has been obtained for a modified RNA polymerase B during sporulation (B′, an enzyme which differs from enzyme B by its later elution from DEAE-sephadex, higher salt sensitivity and template-specifity, but is amatoxin sensitive) and for certain aged strains of *Physarum* (A′), a B-type enzyme that is no longer sensitive to amatoxin) (36). The activity of soluble RNA polymerase depends on the template; double-stranded DNA (native DNA which has been treated with S_1 nuclease) is the most inactive template for the solubilized RNA polymerase B from *Physarum,* as compared with native DNA or denatured DNA, yielding relative incorporation rates of 0.03:0.3:1, respectively. Furthermore, soluble RNA polymerases preferentially transcribe synthetic templates, which are rich in dC or are incompletely double-stranded structures [poly(dA) × oligo(dT) is 20-fold more active than the double-stranded molecule poly(dA) × poly(dT)] (37). The activity of the soluble enzymes can be altered by the addition of endogenous factors: (a) an organic polyphosphate selectively and reversibly inactivates RNA polymerase A of *Physarum* by preventing *in vitro* initiation of transcription. This factor accumulates during starvation in the nucleolus, which may indicate a control mechanism of the restriction of rRNA synthesis during spherulation of *Physarum* (38), (b) a protein factor has been obtained from plasmodia, as well as from isolated nuclei, which stimulates predominantly RNA polymerase B from *Physarum.* This factor does not increase the initiation frequency of *in vitro* transcription, but it leads to generally larger *in vitro* RNA products and may resemble the elongation factors described in other systems. Interestingly, this factor cannot be detected in spherules (39).

b. Endogenous RNA Polymerases. At least a fraction of enzymes A and B is retained in isolated nuclei (40). The resistance of *in vitro* transcription in isolated nucleoli to amatoxin, as well as the molecular size distribution of *in vitro* products, clearly indicates that RNA polymerase A is responsible for the transcription of ribosomal genes of *Physarum* (K. E. Davies, Oxford and A. Hildebrandt, Konstanz, personal communication).

There is no convincing evidence at present for *de novo* initiation of transcription in isolated nuclei of *Physarum,* and this holds true for most other systems (41). Nevertheless, circumstantial evidence allows us to discriminate four different states of nuclear RNA polymerase B, the enzyme assumed to produce HnRNA (Hildebrandt and Sauer, manuscript submitted): (a) free molecules (approximately 80% of the total soluble RNA polymerase B activity, which is released during nuclear isolation), (b) loosely, and (c) tightly bound molecules (15–20%, which are released with 0.5 *M* or 1.5 *M* NaCl, respectively, and (d) "engaged" molecules (0–5% according to physiological conditions), resistant to salt extractions; up to 5×10^4 molecules, according to a preliminary estimate. The main observation from these experiments is the fact that nuclei contain an abundance of free RNA polymerase B (42). The engaged molecules are believed to represent transcription complexes active *in vivo.* Their number can be increased at the expense of the tightly bound enzyme fraction by an incubation of isolated nuclei with a detergent (Triton X 100) *in vitro.* Such transition between bound and putative transcribing molecules may provide for a mechanism for transcription control, as discussed below.

Recently a new method has been devised to prepare "native" chromatin from *Physarum* nuclei lysed by an incubation in lysolecithin. This chromatin can yield *in vitro* RNA transcripts of 10–30 S with endogenous RNA polymerases and also with exogenously supplied homologous RNA polymerase B. *In vitro* initiation of transcription works best if the chromatin has been prepared from nuclei isolated in S phase (unpublished results by C. Schicker in our laboratory).

3. RNA Classes in *Physarum*

RNA of *Physarum* has been analyzed by most available methods. Some problems have arisen due to high nucleolytic activities in the plasmodium and the compact structure of the nucleolus. Nevertheless, there is no doubt that *Physarum* contains the typical RNA classes of eukaryotic cells: tRNA, rRNA, mRNA, and HnRNA (see 10 for review). Small nuclear (sn) RNA's have not yet been described in *Physarum.* tRNA's are stable molecules that are present in a constant proportion to rRNA (43).

Ribosomes contain 26 S and 19 S rRNA (44), as well as 5.8 S and 5 S rRNA's. The larger molecule (5.8 S) is part of the ribosomal cistron transcript; the 5 S RNA is not (L. Hall, personal communication). Ribosomal RNA's are stable molecules that are derived through processing of a nucleolar precursor molecule (approximately 40 S) (45), as in other cells. Stationary growth phase plasmodia contain less ribosomal RNA, which is probably explained by an increased RNAase activity and by the reduced synthetic rate (46).

Polysomes contain RNA other than rRNA which has been purified by affinity chromatography [oligo(dT) cellulose and poly(U) Sepharose] and velocity gradients (47). This RNA contains mRNA as is shown by its activity in *in vitro* translation systems: (a) the poly(A) plus RNA (5–20 S, mean value 10 S, approximately 1% of total RNA) directs protein synthesis in the wheat germ system, (A. Baeckmann, unpublished results); (b) translation of histones has been observed in the reticulocyte system supplemented with poly(A) minus RNA (D. Gallwitz and K. Scheller, unpublished results).

There is evidence for processing of cytoplasmic RNA from somewhat larger molecules synthesized in the nucleus. (a) Nuclear poly(A) plus RNA contain fractions up to 30 S (mean value 10–15 S) (18). (b) Endogenous RNA polymerase B in nuclei leads to the synthesis of up to 30 S RNA, as analyzed under denaturating conditions, although most RNA (80%) sediments with 10 S (A. Hildebrandt and H. W. Sauer, manuscript submitted). (c) Contrary to polysomal poly(A) plus RNA, 5–10% of nuclear poly(A) plus RNA consists of putative double-stranded structures, (possibly transcripts of palindromic DNA sequences), and of up to 25% fast hybridizing base sequences. The remaining RNA (approximately 65–70%) hybridizes with unique DNA sequences, as does all polysomal poly(A) plus RNA (48). (d) At near-saturation conditions, hybridization values indicate that about 10–20% of the DNA is transcribed, whereas 1–2% of DNA sequences are present in cytoplasmic RNA. As a rough approximation, cytoplasmic mRNA may contain transcripts from 5000 genes; (e) Pulse-chase experiments have suggested that about 50% of the nuclear poly(A) plus RNA leave the nucleus within 10–15 minutes, whereas the other half is lost from the nucleus within 6–24 hours. Therefore, a fraction of the RNA transcripts might be stored in the nucleus and translated at a later time. As an estimate, about 10% of nuclear poly(A) RNA reaches the cytoplasm (49). These kinetics data are rather uncertain due to possible fluctuations in isotope uptake, inefficient chase, and reutilization of degradation products. Consequently, an isotope dilution technique has been used to analyze transcription of stable RNA's (50). The degree of stability, or instability of mRNA of *Physarum* is not known, and guesses run from 30 minutes to 3 to 6 hours up to three generation times. Therefore, other mechanisms than RNA transcription may exist to control gene expression in the cell cycle.

However, during the transition from growth to differentiation, the changes in synthesis rates of RNA classes and of endogenous RNA polymerase activities are indications of transcription control mechanisms in *Physarum*.

In summary, our discussion of the transcription machinery of *Physarum* has shown that (a) the templates resemble DNA of higher cells, except for the organization of ribosomal cistrons, (b) RNA polymerases, and (c) RNA products are typical for other eukaryotic systems. We have mentioned some evidence for regulatory mechanisms of gene expression which may operate during differentiation, and we shall now turn to the cell cycle.

In Vivo Transcription in the Cell Cycle of *Physarum*

An early indication for variable transcription in the cell cycle has been a biphasic pattern of uridine incorporation, suggesting two broad peaks, roughly correlating with S phase and with G_2 phase, separated by a valley at the end of S phase (51). Small differences in the base composition of RNA synthesized at various points of the cell cycle have been interpreted to indicate transcription of DNA-like RNA in S phase and rRNA in G_2 phase (52, but see 10). However, radioactivity profiles obtained with sucrose gradients, and polyacrylamide gels have not revealed significant differences in the total RNA for any cell cycle stage (53). A recent hybridization experiment with long incubation periods, in order to saturate at least some of the unique DNA sequences, did indicate a higher heterogeneity of the RNA made in S phase, as compared with RNA from G_2 phase (54). Furthermore, the analysis of poly(A) plus RNA synthesis and content has shown high values in S phase, especially in the early half of S phase, in comparison with RNA from later points of the cycle (by a factor of 2–4) (47). Therefore, poly(A) plus RNA transcription seems to be high in S phase, but does not show a biphasic pattern. A further distinction of nuclear poly(A) plus RNA is the finding that the molecules made in S phase contain less redundant base sequences than those made in G_2 phase (48). If we accept the interpretations given above of (1) early replication of euchromatin, and (2) repeated base sequences are at the 5′ end of nuclear RNA molecules, then, at least to some extent, different portions of the DNA are transcribed in S phase, as opposed to G_2 phase.

From all incorporation data, it seems clear that most RNA synthesized at any time in the cell cycle is ribosomal RNA. Continuous rRNA synthesis has also been shown with the isotope dilution technique mentioned above (50). This investigation has further indicated that (a) no decrease in rRNA synthesis at midcycle; (b) an increase up to fivefold toward the end of the cycle; and (c) a cessation of synthesis in mitosis. Although replication of ribosomal cistrons occurs in G_2 phase, this finding cannot be explained by gene dosage alone. A follow-up study has confirmed these results for tRNA and established a strict correlation of transcription of these two RNA classes in *Physarum* (K. Fink and G. Turnock, personal communication).

The interruption of rRNA synthesis at mitosis is consistent with an earlier autoradiographic observation (55). We have confirmed this result by the analysis of RNA, labeled for 1, 2, or 4 minutes at distinct points around mitosis; at metaphase (for 5–8 minutes), the incorporation into poly(A) plus RNA, large poly(A) minus RNA, and small poly(A) minus RNA (soluble in cetyltrimethylammonium bromide) is inhibited by 95%, 90%, or 50%, respectively (R. Wick, unpublished).

In summary, the *in vivo* analyses indicate fluctuations in the cell cycle of *Physarum* from almost zero in metaphase to maximum transcription of poly(A) plus RNA in early S phase and rRNA, plus tRNA, in late G_2 phase. Furthermore, nuclear RNA synthesized in S phase or G_2 phase differs in hybridization behavior and content of repeated base sequences.

C. *In Vitro* Transcription in the Cell Cycle of *Physarum*

First experiments with isolated nuclei have shown a biphasic pattern of *in vitro* UMP incorporation during the cell cycle by endogenous RNA polymerase similar to that of *in vivo* incorporation (56). The *in vitro* RNA products were of low molecular weight (<4 S), in contrast to the *in vivo* products. A variable incorporation rate has also been found in a following study of *in vitro* transcription with isolated nuclei from a strain of *Physarum,* with exceptionally large nucleoli, under somewhat different ionic conditions (40). In G_2 phase of the cell cycle, the rate of UMP incorporation was about half of that seen in S phase. However, in the presence of 50 m*M* ammonium sulfate, the low endogenous RNA polymerase activity of mid-G_2-phase nuclei was stimulated to the S phase level yielding, once again, a biphasic pattern for the cell cycle. Important new information has come from the differential inhibition on *in vitro* transcription by amatoxin, indicating an activity of RNA polymerase B only in S phase, while RNA polymerase A was equally active in S and G_2 nuclei. After stimulating G_2 phase nuclei with salt (50 m*M* AS), RNA polymerase B activity has appeared in these nuclei, as deduced from partial sensitivity to amatoxin (40).

A reinvestigation of isolated nuclei containing normal size nucleoli under optimum *in vitro* conditions for either RNA polymerase A (0.1 *M* KCl) or RNA polymerase B (0.45 *M* KCl) has indicated that RNA polymerase A is about equally active throughout the cell cycle (A. Hildebrandt, unpublished). Under these conditions, RNA polymerase B activity can be detected at any point of S and G_2 phase, which is consistent with poly(A), plus RNA synthesis throughout interphase. The same results are obtained when nuclei were preextracted to test for the salt-resistant "engaged" enzyme fraction. In these experiments, endogenous enzyme B elongates RNA chains of predominantly mRNA size. An approximately twofold higher activity of endogenous RNA polymerase B has been observed in S phase, as compared to G_2 phase.

However, preliminary observations with detergent-treated nuclei indicate equal activity of enzyme B in S and G_2 phase (A. Hildebrandt and W. D. Grant, personal communication). Together with the *in vitro* stimulation of enzyme B in G_2 phase nuclei at a moderate salt concentration (40) and the decrease of poly(A) plus RNA synthesis in G_2 phase *in vivo* (47) (see above), these experiments might indicate that the capacity of mRNA transcription is reduced *in vivo* by "silencing" transcribing RNA polymerase B molecules. This notion is constant with the appearance of putative "transcription blockers" in the supernatants of the respective nuclear preparations after treatment with detergent (H. Hildebrandt and H. W. Sauer, manuscript submitted).

While most transcription assays with nuclei were done in S or G_2 phase, it is of great importance to know whether metaphase chromosomes of *Physarum* contain RNA polymerase B. Experiments with metaphase nuclei have not revealed any activity of that enzyme. Test were done (a) under optimum ionic conditions, (b) after detergent stimulation, and (c) under conditions which yield "run off" enzyme (A. Hildebrandt, unpublished). This finding, together with the results of *in vivo* transcription, might indicate that for each cell cycle a new transcription program has to be established, beginning in metaphase.

The levels of the two RNA polymerases, determined as solubilized enzymes A and B, do not change over the cell cycle (42). This is consistent with the slow turnover of these proteins and the abundant pools of free RNA polymerase molecules (35). However, the stimulatory factor of RNA polymerase activity shows remarkable fluctuations. A two- to threefold higher amount of this protein seems to be present for about 30 minutes in mid-S phase (1 hour after mitosis) (39). It is suggestive that this factor, which stimulates transcription at a stage after initiation, might play a regulative role *in vivo*, since a high rate of synthesis of poly(A) plus RNA occurs in S phase (47). At present, we have no evidence that the endogenous inhibitor of RNA polymerase A fluctuates during the cell cycle, provided that exponential growth is maintained. Therefore, the low level of *in vivo* rRNA synthesis in S phase, as compared with late G_2 phase (50) cannot be explained by a reduced frequency of initiation on rDNA caused by this inhibitory factor.

D. Protein Synthesis as Possible Marker of Selective Transcription

Although there seems to be some variation in the rate of synthesis and content of poly(A) plus RNA and rRNA, no evidence for selective transcription in the cell cycle has been obtained thus far by the analysis of RNA molecules. Such might be indirectly shown by abrupt changes in protein composition. Possible examples are the fluctuations in the nuclear nonhistone proteins of *Physarum* that occur during starvation and refeeding (57). However, no such changes have been detected during the cell cycle. The same holds true for a large number of proteins that have an affinity for native or denatured DNA (58). In a separate study, proteins were labeled for 1 hour by incubation with [^{35}S]cysteine for each hour of the cell cycle (59). Total protein was fractionated according to solubility in water or acetic acid, subfractionated by hydroxyapatite chromatography and analyzed by autoradiography after SDS polyacrylamide gel electrophoresis. Literally hundreds of labeled protein bands have been detected, but not one significant change of ^{35}S incorporation into these proteins has been observed in the cell cycle of *Physarum*. These results are consistent with analyses of several key enzymes of energy production (glyceraldehyde-phosphate kinase, pyruvate kinase, lactate dehydrogenase, hexokinase, phosphoglucomutase, glucose-6-phosphate dehydrogenase, citrate synthetase, β-hydroxy-acyl-CoA dehydrogenase, unpublished experiments by G. Wegener, in our laboratory), all of which display a linear increase during the cell cycle. Therefore, it can be assumed that most proteins of *Physarum* are synthesized continuously throughout the cell cycle. This might also be indicated by the very similar polysome profiles obtained at any cell cycle stage (60). These observations may indicate that, aside from mitosis and DNA replication, *Physarum* is not physiologically synchronous or, alternately, that enzyme steps seen in other systems, after obtaining synchrony by selection or induction techniques represent perturbations of some kind rather than genuine cell cycle, or more specifically, growth cycle markers.

However, it has been concluded from experiments involving actinomycin D that one enzyme step (RNase, 61) and one enzyme peak (thymidine kinase, 13, 62) may represent transcription events in mid- or late G_2 phase. Therefore, more selective methods have to be employed before selective transcription can be dismissed as a control mechanism of the cell cycle of *Physarum*.

E. Correlations between Transcription and Replication in *Physarum*

The first evidence for an interdependence of RNA and DNA synthesis in *Physarum* has been provided by Rao and Gontcharoff (63). They have shown an inhibition of uridine incorporation when DNA synthesis was inhibited by FUdR. These observations have been confirmed in experiments in which replication was blocked by hydroxyurea or cycloheximide (64). As controls have shown, the application of these inhibitors in G_2 phase has no significant effect on RNA synthesis. The degree of inhibition was highest (up to 70%) in early S phase and decreased in mid- and late S phase (50% and 25%, respectively). In these experiments, incorporation of radioactive precursors into acid precipitable material was measured. In further experiments, RNA, labeled during blocked or normal S phase, has been fractionated according to poly(A) content and molecular weight. All RNA classes are inhibited, but poly(A) plus RNA is most severely affected (up to 60% in early S phase). These incorporation data can also be explained by a selective effect of the drugs on precursor pools in S phase. Evidence for a true inhibition of RNA transcription comes from hybridization of DNA with RNA samples of equal radioactivity. Our results indicate that, especially poly(A) plus RNA, but also nonribosomal poly(A) minus RNA, represent transcription products of less complexity, if RNA preparations from S phase and from blocked S phase are compared. Additional results from hybridization in DNA excess have indicated an abnormal base composition of poly(A) plus RNA made during blocked S phase, which leads to a rapid resistance to RNase in the annealing reaction (64). These results are obtained with RNA transcribed in the presence of drugs that indirectly affect DNA replication by either interfering with precursors or with protein synthesis. The well-documented sequential replication of DNA in *Physarum* makes a more direct test on the correlation of DNA replication and transcription possible. In these experiments, RNA samples of equal radioactivity from plasmodia labeled for 20 or 40 minutes in S phase or G_2 phase have been hybridized to early replicating DNA, (density labeled with BUdR for 30 minutes in the beginning of S phase, and isolated from preparative CsCl gradients—heavy DNA) or to late DNA (which replicates after the first 30 minutes of S phase—light DNA). The results from three independent experiments show that (a) early S phase RNA hybridizes better with early than with late DNA (factor 1.6–2), and that (b) G_2 phase RNA hybridizes about equally well with early or late DNA [approximately 0.5% for poly(A) minus RNA and 2–4% for poly(A) plus RNA]. Significantly, poly(A) plus RNA from cultures blocked in early S phase clearly hybridized less to early DNA than to controls (3%, 6%, 2% versus 6%, 13%, 16%) (49). All these experiments indicate that at least some portion of early replicating DNA seems to be transcribed during or shortly after its replication. This is in agreement with a 70% reduction of endogenous RNA polymerase B in nuclei isolated from blocked S phase (A. Hildebrandt, unpublished)

The inhibition of extractable RNA molecules has never been quite as high as that seen by isotope incorporation alone. One possible interpretation for this difference (of approximately 10%) may be a requirement of RNA synthesis for replication, which is consistent with the fact that the small nascent DNA pieces contain RNA (22). Therefore, there may exist a relation both ways between RNA synthesis and DNA synthesis.

We have mentioned the evidence that protein synthesis is essential for DNA replication. We may also ask whether some protein synthesis depends on DNA replication. When we compare amino acid incorporation into total protein during an S phase blocked with hydroxyurea, and an untreated control, no immediate inhibition has been observed (within 3 hours of S phase). A 10% or 30% reduction of protein synthesis has been noted at 2 or 4 hours after S phase, respectively (59). However, incorporation of basic amino acids into histones is inhibited by about 50% during the blocked S phase. So far, we have no evidence (from the effect of actinomycin D, at 250 μg/ml) that histone synthesis depends on concomitant RNA synthesis. Furthermore, a compensation for this inhibition of histone labeling must occur, since the histone content increases disproportionately in plasmodia blocked in DNA synthesis with hydroxyurea (65). From these preliminary experiments, we can conclude that most proteins synthesized during S phase are not affected by the DNA replication, which is in contrast to the results of RNA transcription. Therefore, DNA-replication-related RNA may be the consequence of uncontrolled transcription at maximum capacity in early S phase and serve other functions, if any, besides directing polypeptide synthesis.

IV. A MODEL OF TRANSCRIPTION CONTROL IN THE CELL CYCLE OF *PHYSARUM*

Although no instance can be cited for an essential and specific gene expression event in the cell cycle, there is a very clear demonstration for transcription control, which results (1) in the cessation of transcription in metaphase, and (2) in predominant transcription of poly(A) plus RNA in S phase and rRNA in late G_2 phase. Therefore, the cell cycle of *Physarum* can conveniently be discussed, beginning at metaphase, a point of realignment of the transcription machinery (4). This point may coincide with the complete loss of RNA polymerases (especially enzyme B) and poly(A) RNA from the nucleus following chromatin condensation. Figuratively, in metaphase, a reset button has been pressed, and we postulate that transcription is programmed in S phase. Since intact double-stranded DNA molecules are the inactive templates for eukaryotic RNA polymerases, whereas (dC)-rich and "gapped" structures (like the discontinuously replicating DNA molecules) are good templates (37), and since DNA in S phase nuclei seems to be more exposed (i.e., accessible to DNase, fast migration in an electric field, see above) and since "native" S phase chromatin (see above) can initiate *in vitro* transcription with RNA polymerase B from *Physarum*, we assume that chromatin is recharged from the large pool of free RNA polymerase molecules (35) that bind to the replicating chromatin structure as one component of the nonhistone protein fraction. Some of these enzyme molecules succeed in immediate initiation of transcription. It is this portion of RNA polymerase that accounts for the "replication–transcription coupling" demonstrated above, especially in early S phase. It is worth mentioning that a correlation of DNA and RNA synthesis has been observed in mammalian cells (66, 67). It is evident from the sequential replication of DNA in *Physarum* that RNA transcription in the nucleus might also be programmed sequentially. This notion is reminiscent of the "linear-reading" hypothesis of Halvorson (68). A further speculation would explain how a change in the sequence of replication during development might program a cell for different transcripts, and, hence, for possibly different functions, which bears on the "quantal cell cycle" concept of Holtzer (69). Such speculation can be tested during *Physarum* sporulation and encystment.

The high rate of RNA synthesis in S phase is not only an expression of initiation frequency, but may also, particularly in mid-S phase of *Physarum*, result from the action of an elongation factor (see above) that leads to the production of larger molecules, either by a higher rate of elongation or by a stabilization of transcription complexes. Such putative mechanism is reminiscent of an "anti-termination" control of transcription.

Since much of the RNA made in S phase remains in the nucleus, it may have regulatory functions in gene expression. Other explanations are that this RNA may (a) represent a pool for some later (G_2 phase) protein synthesis, (b) serve as unessential substrate for RNases, or (c) function in some exotic way, like counting minutes until the next mitosis.

It is a reasonable explanation, for the transcription of poly(A) plus RNA in G_2 phase, to assume that RNA polymerase B molecules select certain sites on the DNA and produce RNA. However, an alternative explanation may be that transcription of DNA in G_2 phase is also determined by those RNA polymerase molecules which have bound to the chromatin during its replication and which have rested in a preinitiated, or repressed state (presumably at stretches of repeated base sequences), and become active by a process that leads to proper initiation of transcription. Some evidence for "resting" RNA polymerase molecules in *Physarum* is provided by the different states of free, bound, and engaged enzyme defined above, and the distinction of bound endogenous enzyme from free soluble enzyme by its high salt requirement (0.45 *M* NaCl vs. 0.15 *M* NaCl). The most significant indication for "preinitiation–initiation–transition" comes from observations on endogenous RNA polymerase activities of isolated nuclei; the proportion of salt-resistant engaged enzyme B can be increased by an incubation with a detergent and rNTP precursor at the expense of a bound enzyme fraction that is soluble in salt (A. Hildebrandt and H. W. Sauer, manuscript submitted). These experiments allow us to discriminate between potential and actual RNA polymerase activity and answer the question whether the decrease in hnRNA transcription during G_2 phase of *Physarum* is caused by a release of RNA polymerase B molecules from the template or by an inactivation of the bound enzyme by the putative detergent-sensitive nonhistone "blockers."

These speculations hold best for fast-cycling cells. However, since it is not known whether RNA polymerase molecules, at termination of transcription, leave the template, it may be that preinitiated RNA polymerase molecules, attached sequentially to the chromatin during DNA replication (and maturation due to excision of primer RNA, DNA repair, and ligation) may determine the capacity for transcription, in resting cells, too. In this respect it is noteworthy (a) that giant chromosomes, after a short incubation at 0.5 *M* salt, show a distinct pattern of puff induction, which has been interpreted as an activation of resting RNA polymerase B molecules (70), (b) that other cells contain free and engaged RNA polymerase at the chromatin level (71) and (c) that a stimulation of bound RNA polymerase with detergent (72) and of soluble RNA polymerase by protein factors (73) is well established in other systems.

In order to generalize this concept, it will be interesting to see whether the chromatin of mammalian cells, which show a correlation of replication and transcription in experiments with hydroxyurea (66), is "cleaned" of RNA polymerase molecules. If so, it will be even more important to learn if this happens at the G_2 M or G_1 S transition points of typical cell cycles.

V. CONCLUSIONS

We have discussed some of the work, which relates to gene expression control mechanisms, that has been done with *Physarum*. The experimental results pertaining to the regulation of nuclear mitotic divisions have not been included.

We feel that (a) the good mitotic synchrony of *Physarum* provides for an excellent experimental system for cell biologists; (b) the available evidence is indicative of, but far from conclusive of, a causal interrelation of gene expression and induction of mitosis; (c) the lack of a main section of a typical cell cycle (G_1 phase) is a virtue that possibly enables us to discriminate control mechanisms of cell proliferation from those operating in the basic cell cycle, and (d) more meaningful experiments can be done with *Physarum.*

ACKNOWLEDGMENTS

I wish to thank Dr. Armin Hildebrandt for his collaboration and valuable discussions during the past 5 years; the many friends and colleagues who offered unpublished information and stimulating thoughts at this year's *Physarum* Conference in Rüttlihubelbad, Switzerland; and the Deutsche Forschungsgemeinschaft for financial support for the work done in my laboratory.

REFERENCES

1. Prescott, D. M., *Adv. Genet.* **18,** 99 (1976).
2. von Stosch, H.-A., *in* "Handbuch der Pflanzenphysiologie" (W. Ruhland, ed.), Vol. 15, Part I, p. 641. Springer-Verlag, Berlin and New York, 1965.
3. Alexopoulus, C. J., *in* "The Fungi: An Advanced Treatise" (G. C. Ainsworth and A. S. Sussman, eds.), Vol. 2, p. 211. Academic Press, New York, 1966.
4. Rusch, H. P., *Adv. Cell Biol.* **1,** 297 (1970).
5. Sauer, H. W., *Symp. Soc. Gen. Microbiol.* **23,** 375 (1973).
6. Schiebel, W., *Ber. Dtsch., Bot. Ges.* **86,** 11 (1973).
7. Mohberg, J. *Cell Nucleus* **1,** 187 (1974).
8. Jockusch, J., B.M., *Ber. Dtsch., Bot. Ges.* **86,** 39 (1973).
9. Lestourgeon, W. M., and Wray, W., *in* "Acidic Proteins of the Nucleus (I. L. Cameron and J. R. Jeter, Jr., eds.), p. 59. Academic Press, New York, 1975.
10. Grant, W. D., *Br. Soc. Cell Biol. Symp.* p. 77 (1973).
11. Mitchison, J. M. *Symp. Soc. Gen. Microbiol.* **23,** 198 (1973).
12. Dee, J., *Ber. Dtsch. Bot. Ges.* **86,** 93 (1973).
13. Sachsenmaier, W., von Fournier, D., and Gürtler, K. F., *Biochem. Biophys. Res. Commun.* **27,** 655 (1967).
14. Fouquet, H., Wick, R., Böhme, R., and Sauer, H. W., *Arch. Biochem. Biophys.* **168,** 273 (1975).
15. Bohnert, H. J., *Exp. Cell Res.* **106,** 426 (1977).
16. Molgaard, H. V., Matthews, H. R., and Bradbury, E. M., *Eur. J. Biochem.* **68,** 541 (1976).
17. Bohnert, H. J., Schiller, B., Böhme, R., and Sauer, H. W., *Eur. J. Biochem.* **57,** 361 (1975).
18. Karrer, K. M., and Gall, J., *J. Mol. Biol.* **104,** 421 (1976).
19. Hourcade, D., Dressler, D., and Wolfson, J., *Proc. Natl. Acad. Sci. U.S.A.* **70,** 2926 (1973).
20. Fouquet, H., Bierweiler, B., and Sauer, H. W., *Eur. J. Biochem.* **44,** 407 (1974).
21. Nygaard, O. F., Guttes, S., and Rusch, H. P., *Biochim. Biophys. Acta* **38,** 298 (1960).

22. Waqar, M. A., and Huberman, J. A., *Biochim. Biophys. Acta* **383,** 410 (1975).
23. Funderud, S., and Haugli, F., *Nucleic Acids Res.* **2,** 214 (1975).
24. Wille, J. E., Jr., and Kauffmann, S. A., *Biochim. Biophys. Acta* **407,** 158 (1975).
25. Braun, R., and Wili, H., *Experientia* **27,** 1412 (1970).
26. Fouquet, H., and Sauer, H. W., *FEBS Lett.* **61,** 234 (1976).
27. Mohberg, J., and Rusch, H. P., *Arch. Biochem. Biophys.* **134,** 577 (1969).
28. Vogt, M., and Braun, R., *FEBS Lett.* **64,** 190 (1974).
29. Jockusch, J. B. M., and Walker, J. O., *Eur. J. Biochem.* **48,** 417 (1974).
30. Hildebrandt, A., and Sauer, H. W., *FEBS Lett.* **35,** 41 (1973).
31. Burgess, A. B., and Burgess, R. R., *Proc. Natl. Acad. Sci. U.S.A.* **71,** 1174 (1974).
32. Weaver, R. F., *Arch. Biochem. Biophys.* **172,** 470 (1976).
33. Hildebrandt, A., and Sauer, H. W., *Biochim. Biophys. Acta* **425,** 316 (1976).
34. Grant, D., and Russel, T., and Poulter, M. J., *Mol. Biol.* **73,** 439 (1973).
35. Hildebrandt, A., and Sauer, H. W., *Wilhelm Roux' Entwicklungsmech. Arch. Org.* **180,** 149 (1976).
36. Hildebrandt, A., and Sauer, H. W., *Verh. Dtsch. Zool. Ges.* (W. Rathmayer, ed.), Gustav Fischer Verlag, Stuttgart p. 24, (1976).
37. Hildebrandt, A., and Sauer, H. W. *Arch. Biochem. Biophys.* **176,** 214 (1976).
38. Hildebrandt, A., and Sauer, H. W., *Biochem. Biophys. Res. Commun.* **74,** 466 (1977).
39. Ernst, G., and Sauer, H. W., *Eur. J. Biochem.* **74,** 253 (1977).
40. Grant, W. D., *Eur. J. Biochem.* **29,** 94 (1972).
41. Chambon, P., *Annu. Rev. Biochem.* **44,** 613 (1975).
42. Hildebrandt, A., and Sauer, H. W., *Biochim. Biophys. Acta* **425,** 316 (1976).
43. Melera, P. W., and Rusch, H. P. *Biochemistry* **12,** 1307 (1973).
44. Melera, P. W., Chet, I., and Rusch, H. P. *Biochim. Biophys. Acta* **209,** 569 (1970).
45. Melera, P. W., and Rusch, H. P., *Exp. Cell Res.* **82,** 197 (1973).
46. Hildebrandt, A., Fouquet, H., Wick, R., and Sauer, H. W., *Verh. Dtsch. Zool. Ges.* p. 150 (1975).
47. Fouquet, F., Böhme, R., Sauer, H. W., and Braun, R., *Biochim. Biophys. Acta* **353,** 313 (1974).
48. Fouquet, H., and Sauer, H. W., *Nature* (*London*) **255,** 253 (1975).
49. Wick, R., Ph.D. Thesis, University of Konstanz, Germany (1976).
50. Hall, L., and Turnock, G., *Eur. J. Biochem.* **62,** 471 (1976).
51. Mittermayer, C., Braun, R., and Rusch, W. P., *Biochim. Biophys. Acta* **91,** 399 (1964).
52. Cummins, J. E., Weisfeld, G. E., and Rusch, H. P., *Biochim. Biophys. Acta* **129,** 240 (1966).
53. Zellweger, A., and Braun, R., *Exp. Cell Res.* **65,** 413 (1971).
54. Fouquet, H., and Braun, R., *FEBS Lett.* **38,** 181 (1974).
55. Kessler, D., *Exp. Cell Res.* **45,** 676 (1967).
56. Mittermayer, C., Braun, R., and Rusch, H. P., *Biochim. Biophys. Acta* **114,** 536 (1966).
57. Lestourgeon, W. M., and Rusch, H. P., *Science* **174,** 1233 (1971).
58. Magun, B. E., Burgess, R. R., and Rusch, H. P., *Arch. Biochem. Biophys.* **170,** 49 (1975).
59. Ernst, G., Ph.D. Thesis, University of Konstanz, Germany (1976).
60. Brewer, E. N., *Biochim. Biophys. Acta* **277,** 639 (1972).
61. Braun, R., and Behrens, K., *Biochim. Biophys. Acta* **195,** 87 (1969).
62. Sachsenmaier, W., and Ives, D. H., *Biochem. Z.* **343,** 399 (1965).
63. Rao, B., and Gontcharoff, M., *Exp. Cell Res.* **56,** 269 (1969).
64. Fouquet, H., Böhme, R., Wick, R., and Sauer, H. W., *J. Cell Sci.* **18,** 27 (1975).
65. Mohberg, J., and Rusch, H. P., *J. Bacteriol.* **97,** 1411 (1969).

66. Klevecz, R. R., and Stubblefield, E., *J. Exp. Zool.* **165,** 259 (1967).
67. Pfeiffer, S. E., and Tolmach, L. J., *J. Cell. Physiol.* **71,** 77 (1968).
68. Halvorson, H. O., Carter, B. L. A., and Tauro, P., *Adv. Microbial Physiol.* **6,** 47 (1971).
69. Holtzer, H., Sanger, J. W., Ishikawa, H., and Strahs, K., *Cold Spring Harbor Symp. Quant. Biol.* **37,** 549 (1973).
70. Hameister, H., and Pelling, C., *Nachr. Akad. Wiss. Göttingen* **11,** 40 (1975).
71. Yu, L.-F., *Nature (London)* **251,** 344 (1974).
72. Chesterton, C. J., Coupar, B. E. H., Butterworth, P. H. W., Buss, J., and Green, M. H., *Eur. J. Biochem.* **57,** 79 (1976).
73. Seifart, K. H., Juhasz, P. P., and Benecke, B. J., *Eur. J. Biochem.* **33,** 181 (1973).

9

Regulation of Protein Synthesis during the Cell Cycle in *Chlamydomonas reinhardi*

Stephen H. Howell and D. Mona Baumgartel

I. INTRODUCTION

A question about the regulation of protein synthesis during the cell cycle can be framed in one of two ways, depending on your point of view. One can ask what effect the cell cycle has on protein synthesis, or, approaching the question from a different direction, what effects do changes in protein synthesis have on the cell cycle? To decide which is the cart and which is the horse is difficult because both are such all-pervading processes in the cell that each must surely influence the other.

During the cell cycle, the synthesis of a single protein species could be controlled by any component of the genetic apparatus. It is possible that the synthesis of a protein is regulated at the level of transcription, messenger processing and transport, and/or at the level of translation. Furthermore, regulatory controls could be general, involving changes in the overall rate of transcription or translation, or specific in the sense that a particular polypeptide species may be subject to regulatory controls from which others are exempt. Despite the possibility that the synthesis of a protein may be regulated at many different levels during the cell cycle, we will focus our attention on strictly translational controls. This does not oversimplify our task because protein synthesis itself is a highly complex process involving a myriad of components. An appreciation for the intricacies of the process and the properties of translation factors can be had by reading a recent review by Weissbach and Ochoa (1). Also, Lodish (2) has recently reviewed instances where translational controls may be involved in biological processes other than the cell cycle.

Studies of protein synthesis during the cell cycle have not yet revealed detailed changes in the workings of the protein-synthesizing machinery at each cell cycle stage; i.e., they have not shown variations in the activity of individual translation factors at different periods of the cell cycle. Instead, studies have generally followed one of two alternative courses. Either they have examined the overall rates of protein synthesis at various cell cycle stages, while viewing the protein-synthesizing mechanism as a black box, or they have studied the kinetics of accumulation of a single polypeptide (or enzyme) during the cell cycle. Both approaches have yielded interesting results and have opened pathways along which more mechanistic studies can follow.

A. Control of the Synthesis of Single Polypeptide Species in the Cell Cycle

Numerous reports on the stage-specific appearance of periodic enzymes in synchronous cells have shown that the syntheses of many individual polypeptides are independently regulated in the cell cycle. Nearly 30 enzymes in synchronous *Saccaromyces cerevisiae* cultures have been reported to show stepwise increases in activity at various cell cycle stages (see reviews 3, 4). The theories that have been most often advanced to explain the periodic synthesis of these enzymes do not involve translational mechanisms, however. It has been suggested that stage-specific enzyme synthesis in yeast results from sequential transcription or duplication (4) of the genome during the cell cycle.

Until recently, proponents of the point of view that transcriptional events control periodic polypeptide synthesis in synchronous cells were hard pressed to cite examples where cell-cycle-specific transcription of genes could be documented. With the development of techniques to detect RNA transcripts from single genes using copy or cloned DNA hybridization probes, at least one possible case of stage-specific transcription has been found. That is the case of histone synthesis. It has been shown that histone messenger RNA (mRNA) accumulates on polyribosomes during S phase and that this accumulation is tightly coupled to DNA synthesis (5, 6). Stein *et al.* (7) have suggested that this regulation may be transcriptional and may involve the activation of histone genes during S phase by nonhistone chromosomal proteins. As probes for other mRNA's are developed, more possible examples of stage-specific transcription may come to light.

Examples of the synthesis of periodic enzymes thought to be controlled by purely translational mechanisms are rare. A possible one is the oscillatory synthesis of lactic dehydrogenase during the cell cycle in synchronous Chinese hamster cells as reported by Klevecz (8). He found that in an ongoing cell cycle the stage-specific synthetic pattern of this enzyme did not require RNA synthesis, i.e., was insensitive to actinomycin D. This finding, however, will have to await further confirmation that actinomycin D at such concentrations did, in fact, block the synthesis of lactic deydrogenase mRNA.

Messenger RNA turnover rates have an important bearing on whether strict transcriptional control mechanisms could regulate the periodic synthesis of stage-specific polypeptides. Obviously, a periodically synthesized protein with a long-lived mRNA would be a poor candidate for regulation by strictly transcriptional means. In the best-studied organisms, where both mRNA half-lives and periodic enzyme synthesis have been studied, the possibilities for strict transcriptional control are not clear. Transcriptional regulation during the cell cycle would seem possible in *Saccharomyces cerevisiae,* since the half-life of mRNA is about one-tenth the doubling time (9), and, perhaps in *Schizosaccharomyces pombe,* where the half-life is somewhat longer, about one-third the cell cycle (10). However, in cultured mammalian cells, the half-life of much of the poly(A)-associated RNA is considerably longer, from half to the full doubling time (see review 11). Of course, comparing bulk messenger turnover rates to doubling times would not rule out transcriptional control during the cell cycle if the mRNA's coding for periodically synthesized polypeptides had very different turnover rates than the bulk mRNA.

Accounting for the rate of synthesis of a protein during the cell cycle by strictly transcriptional means can be quite complicated, as illustrated by the example of sucrase and acid phosphatase synthesis in *S. pombe.* Both enzymes accumulate at one linear rate through the first 20% of the cycle and at a greater linear rate beyond that stage (12). (These enzymes do not accumulate by periodic synthesis in the manner we have been discussing.)

Although the accumulation pattern during the cell cycle is nearly identical for both enzymes, the mRNA for sucrase is quite stable, while that for acid phosphatase has a half-life of about 25 minutes (13). Thus, by a strict transcriptional control mechanism, a burst of sucrase mRNA synthesis at the 20% mark in the cell cycle would be required to generate the pattern of sucrase synthesis, and to obtain the pattern of acid phosphatase synthesis there would have to be a constant rate of acid phosphatase mRNA synthesis until the 20% mark and about a doubling of the rate thereafter (13). The possibility that such divergent transcriptional tactics could produce the same enzyme accumulation pattern seems remote and suggests other means of regulation.

It is no less difficult to envision the regulation of individual, periodically synthesized proteins during the cell cycle by translational means. Translational control is not usually thought to be polypeptide specific, since the evidence for mRNA-specific translation factors is not very strong (2). Nonetheless, at different cell cycle stages, the selection of specific mRNA's by strictly translational means is not out of the question and is a point to which we will return later.

B. Control of the Overall Rates of Protein Synthesis in the Cell Cycle

Variations in the overall rate of protein synthesis during the cell cycle were first reported in autoradiographic studies such as those of Prescott and Bender (14). They showed that rate of amino acid incorporation in Chinese hamster and HeLa cells dropped during mitosis. Their observations have been reproduced subsequently, and finding the reason for the slowdown of protein synthesis during mitosis in synchronized mammalian cells has been a dominant theme in other, later studies (15–19).

The suspected step in controlling protein synthesis during the cell cycle is polypeptide chain initiation, since it appears that translational control in eukaryotic cells involves primarily regulation of chain initiation and not chain elongation or termination (2). That generalization is based on few examples, such as in the reticulocyte system, where it has been shown that polypeptides which are synthesized at different rates, such as α- and β-globins and other minor proteins, are initiated, but not elongated, at the different rates (20, 21). Initiation of translation involves the formation of a mRNA-Met-tRNA$_f$-40 S ribosome complex, and the rate of initiation could be governed by any of the components of the complex. The availability of mRNA for forming initiation complexes could be controlled by transcription, sequestration in the cell, complexing with protein, modification [such as addition of a 3′-poly(A) tail or 5′-7 MeG cap], or degradation through processing or "normal use." However, there is not yet any data on what factors control the formation of initiation complexes at various cell cycle stages—only the information that initiation rates change during the cell cycle.

Fan and Penman (15) showed that polypeptide chain initiation rates varied during the cell cycle in Chinese hamster ovary cells. They found that the rate of initiation dropped threefold in mitotic cells based on the following observations. First, the rate of polypeptide elongation, which they called the transit time (roughly, the time to translate an mRNA of average length), was the same in mitosis-arrested and interphase cells. Second, the average size classes of polyribosomes and the sizes of polypeptides synthesized were about the same in mitotic and interphase cells. Third, and the contrasting condition, the percent ribosomes in polyribosomes in mitotic cells was one-third the level in interphase cells. Thus, it appeared that protein synthesis was slower in mitotic cells because a smaller number of ribosomes were actively engaged in protein synthesis. However, because the rate of chain elongation was the same at these two cell cycle stages, when a ribosome intitiated protein synthesis in a mitotic cell, it was able to move along a messenger RNA at the same rate as in an interphase cell. The build-up of a nontranslating ribosome (subunit) pool in mitotic cells was found by Fan and Penman (15) to be due to a limitation in forming initiation complexes, and not to some inability of the nontranslating ribosomes to function. This was shown by the fact that ribosomes in mitotic cells could be mobilized for protein synthesis by slowing down the rate of chain elongation with low doses of cycloheximide.

Fan and Penman (15) also concluded from the cycloheximide-treatment experiment that the amount of translatable mRNA in mitotic cells did not limit translation initiation. They deduced this from the fact that the size and quantitity of polyribosomes in cycloheximide-treated mitotic cells were nearly the same as in interphase cells. Hence, mitotic and interphase cells seemed to possess about the same quantity of translatable mRNA, i.e., RNA that could be loaded with ribosomes to form polyribosomes. Fan and Penman (15) clearly implicated some factor (soluble or ribosomal) other than mRNA as the limitation to chain initiation in mammalian mitotic cells.

Different conclusions had been reached in earlier experiments by Salb and Marcus (16) studying *in vitro* protein synthesis in extracts derived from vinblastine-arrested HeLa cells. They found that, although supernatant (S 100) fractions from interphase and mitotic cells were equally capable of poly(U)-directed phenylalanine incorporation in a reconstructed system, ribosomes from the two different cell cycle stages were not. Ribosomes derived from mitotic cells were definitely less efficient in an *in vitro* system, but could be reactivated to interphase levels by trypsin treatment. Thus, like

Fan and Penman (15), they concluded that not all ribosomes in mitotic cells were active; however, Salb and Marcus (16) suggested, in contrast, that the nontranslating ribosomes had been inactivated by a trypsin-sensitive inhibitor. In retrospect, it was probably not necessary for Salb and Marcus to conclude that trypsin activated only the nontranslating ribosomes in mitotic cells. They might have proposed as an alternative that trypsin stimulated the activity of all ribosomes derived from mitotic cells.

Johnson and Holland (17) obtained results nearly opposite to those of Salb and Marcus (16). They found almost no difference in the rate of poly(U)-directed phenylalanine incorporation by extracts from interphase and vinblastine-arrested HeLa cells. They also found that interphase and mitosis-arrested cells supported poliovirus multiplication to about the same extent, and so concluded that there was no difference in the protein-synthesizing capacity required for poliovirus multiplication at either cell cycle stage. They argued that protein synthesis in mitotic cells was limited by mRNA availability, and not by some temporary impairment of the translational machinery.

Two later observations, those of Steward *et al.* (18), studying Chinese hamster cells and of Hodge, Robbins, and Scharff (19), studying HeLa cells, indicated that mRNA probably did not limit protein synthesis during mitosis. Both of these studies showed that polyribosome disassembly occurred at mitosis and that the subsequent reassembly following mitosis did not require *de novo* RNA synthesis, i.e., was not blocked by actinomycin D. It was suggested, therefore, that untranslated mRNA is found in mitotic cells and that it is translated (or becomes available for translation) as the cell moves out of mitosis. This was also the conclusion reached by Fan and Penman (15) in their experiments. Thus, there are arguments on both sides as to whether translatable mRNA is found in mitotic cells. As described earlier, the turnover rate of mRNA in mammalian cells suggests that there is no large-scale destruction of messages at mitosis. A recent study by Fraser and Carter (22) has shown that the quantity of poly(A)-associated RNA in *Saccharomyces cerevisiae* does not decline during mitosis. In *Physarum polycephalum,* mRNA seems to persist through mitosis because polyribosomes do not disaggregate at that time (23). This latter observation is interesting because it suggests that the drop in protein synthesis during mitosis in *Physarum* (24) must not result from a limitation in initiation, which would tend to disaggregate polyribosomes. Instead, it would seem that chain elongation slows during the synchronous mitosis in this uninucleate organism.

So, in addition to the problem of conflicting reports on the translation activities of cells during the cell cycle in the same organism, there are also differences recorded between various organisms. It may be that different organisms or different cells in the same organism have different cell cycle strategies. We hope for simplicity's sake that this is not so. The real conflicts may simply be experimental and due to the fact that separate observations have been made in different laboratories using different cells. What we will discuss next is a complete investigation that we have carried out on a single organism, *Chlamydomonas reinhardi*. We have studied the changes in initiation and elongation rates throughout the cell cycle (and not simply at two different cells cycle stages), and we will compare these changes in the rates of overall protein synthesis to those seen for individual polypeptides.

II. STUDIES WITH *CHLAMYDOMONAS REINHARDI*

Chlamydomonas reinhardi is an ideal subject for cell cycle studies because of the ease with which one can synchronize its cultures and obtain samples representing cells at discrete cell cycle stages. Cultures were synchronized by 12-hour light-12-hour dark cycles, and, under such illumination conditions, cell separation occurred about midway (4–6 hours) during the dark period. Other cell cycle events occurred at regular intervals (see 25). *C. reinhardi* cells contain two different protein-synthesizing systems, one cytoplasmic and the other organellar (mostly chloroplastic). In what follows, we will describe primarily the cell cycle properties of the cytoplasmic system.

A. Control of the Overall Rates of Protein Synthesis in the Cell Cycle

It is difficult to determine changes in the rates of protein synthesis during the cell cycle by measuring the rate of incorporation of labeled amino acids at various cell cycle stages. Incorporation rates may not be true measures of protein synthesis because the specific activity of intracellular precursor pools from which proteins are synthesized may vary at different cell cycle stages. Under such conditions, it may not even be a useful control to measure the specific activity of an intracellular amino acid because the pool used for protein synthesis may be separate or discrete. An example of this problem was described by Howell *et al.* (26) in a study where they showed that the incorporation patterns of $^{35}SO_4^{3-}$ and [^{3}H]arginine during the cell cycle in *C. reinhardi* were quite different from each other, and as we will show here, different from the actual changes in rate of protein synthesis in the cell cycle (27).

To circumvent these difficulties, we measured rate changes in protein synthesis with a technique that is unaffected by changes in precursor pools.

The technique involves determining the independent rates of polypeptide initiation and elongation and then combining these measurements to arrive at the rate of protein synthesis (27). We measured chain elongation rates with a procedure devised by Enger and Tobey (28) for mammalian cells and applied to yeast by Petersen and McLaughlin (29). Basically, it involves determining the rate at which polyribosomes of one size acquire radioactive label in nascent polypeptide chains, as compared to the rate in polyribosomes of another size. The measurement gives an internal comparison of the labeling rates of nascent chains on polyribosomes of different sizes, which presumably draw amino acids from the same precursor pool. By relating these measurements from cells at different cell cycle stages, one can obtain relative changes in the rate of polypeptide chain elongation during the cell cycle.

During a very short pulse-labeling period with an amino acid, the specific activity (cpm in nascent peptide chains/A_{260} in ribosomes) of polyribosomes of different sizes will be about the same, since ribosomes on polyribosomes of any size will synthesize the same small segment of polypeptide during a pulse. With longer pulse-labeling times, the specific activity of increasingly larger polyribosomes will increase to a limit defined by their average nascent chain content, and at this limit, or at labeling saturation, a constant ratio of specific activities of the two classes of polyribosomes ($N = 9/N = 6$) is reached (30). By pulse-labeling cells and measuring this ratio, one can obtain a value for I, the peptide interval time, which is the time interval between the labeling saturation of one size class polyribosome and the next larger class. (At a polyribosome level, it is the time required for a ribosome to complete a peptide interval unit, the amount of peptide translated as a ribosome moves the distance equal to the average center-to-center spacing of ribosomes on a polyribosome.) In practice, one can measure I by pulse-labeling cells for a period less than the time required to saturate either of the size classes of polyribosomes to be compared. In these experiments, we have studied the polyribosomes of size class $N = 9$ and $N = 6$. Polyribosomes in these size classes are almost exclusively cytoplasmic, not chloroplastic, and accumulate nascent chain according to expectations (31). One then uses the ratio of specific activities of the two classes of polyribosomes ($N = 9/N = 6$) to determine I from a plot of the ratio of specific activities versus the number of peptide interval units completed during the pulse. A complete description of this technique is described by Enger and Tobey (28) and by Baumgartel and Howell (27).

The results of such an experiment are found in Fig. 1A, where the ratio of specific activities of polyribosomes $N = 9$ and $N = 6$ are shown following a 30-second [^{3}H] arginine pulse at different cell cycle stages. In Fig. 1B, the corresponding values of I, peptide interval time, have been calculated (27).

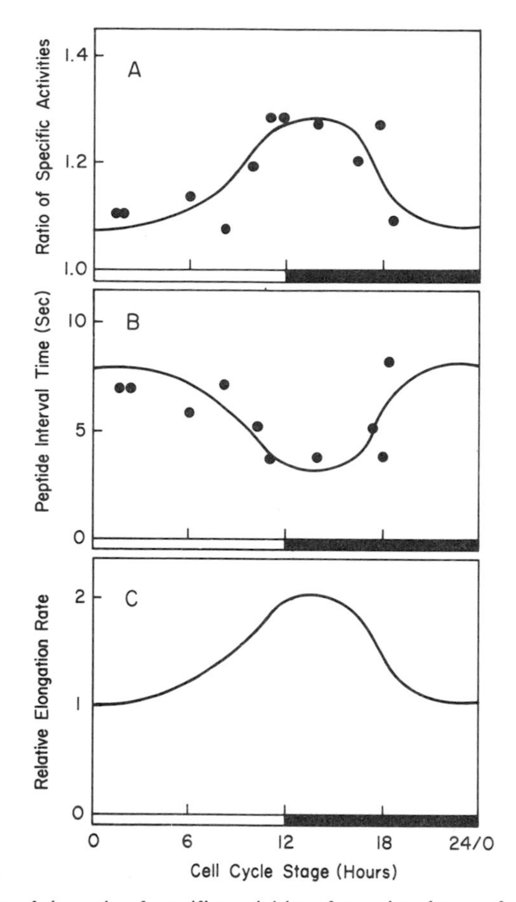

FIG. 1. (A) Plot of the ratio of specific activities of two size classes of polyribosomes from cells pulse labeled at different times in the cell cycle. Cells were labeled for 30 seconds with [^{3}H]-arginine, and the specific activities of polyribosomes $N = 9/N = 6$ were obtained from the sucrose-gradient profiles of extracted polyribosomes. Specific activities ($N = 9/N = 6$) are expressed as cpm in nascent peptide chains/A_{260} in ribosomes. (B) Peptides interval times, I, were calculated according to Baumgartel and Howell (27). (C) Relative elongation rates, r_e, calculated according to Eq. (1). Parameter c has been adjusted to normalize initial rate (time 0) to 1. Twelve-hour light–12-hour dark illumination cycle represented by open and closed bars. Cell division occurred at about 16 hours or 4 hours into the dark period. [From Baumgartel and Howell (27).]

It can be seen that I was not constant throughout the cell cycle but varied from about 8 seconds in the early light phase to about 4 seconds at the beginning of the dark phase. The chain elongation rate, r_e (the number of amino acids added per chain per unit time), is the reciprocal of I or

$$r_e = c/I \tag{1}$$

where c is an undetermined constant, which describes the number of amino acids in a peptide unit. Relative changes in r_e shown in Fig. 1C were about twofold in the cell cycle, and the greatest rates were reached at the time of cell division (which occurs about 4 hours into the dark phase).

To measure changes in chain initiation rates during the cell cycle (defined as the number of chain initiation events *per cell* per unit time), we determined values for P, the number of ribosomes in polyribosomes per cell, at different cell cycle stages. P depends on the balance between the rates of chain initiation and elongation as described by Vassart *et al.* (32) and on the synthesis of new ribosomes. If the rates of chain elongation, r_e, and synthesis of new ribosomes during the cell cycle are known, then P can be used to determine chain initiation rates. Cattolico *et al.* (33) have previously reported on the synthesis of ribosomal RNA during the cell cycle in *C. reinhardi*, and, from their data, the number of ribosomes per cell at various cell cycle stages can be obtained. We have determined the proportion of ribosomes in polyribosomes at different cell cycle stages as shown in Fig. 2, and, with the data of Cattolico *et al.* (33), we can calculate P at various cell cycle stages. In Fig. 2, it can be seen that the proportion of ribosomes in polyribosomes (total chloroplastic and cytoplasmic in Fig. 2A and cytoplasmic in Fig. 2B) was fairly constant throughout the light period, but, at the beginning of the dark, polyribosomes began to disaggregate and continued to do so until the end of the dark period. Polyribosomes were rapidly formed again at the beginning of the light period, about 6 hours after the time of cell separation.

P is related to the initiation rate, r_i (number of chain initiation events per cell per unit time) according to Eq. (27):

$$P = m(r_i/r_e) \tag{2}$$

where m is a parameter describing the average size of the mRNA population being translated. Thus, P increases with r_i at constant r_e and declines with increasing r_e at constant r_i. We suggest that m was relatively invariant during the cell cycle in *C. reinhardi* because the size distribution profiles of polyribosomes were not significantly different at various cell cycle stages (27).

We have calculated the relative changes in r_i during the cell cycle according to Eq. (2) and have plotted these changes in Fig. 3. The figure shows that r_i fluctuated more than any other translational parameter described so far in the *C. reinhardi* cell cycle—more than 25-fold. At the beginning of the light period, the increase in r_i led to a rapid recruitment of ribosomes in polyribosomes at a time when r_e (see Fig. 1C) and the number of ribosomes per cell were low. Continuing into the light period, r_i steadily increased to maintain a remarkable regulation of the proportion of ribosomes in

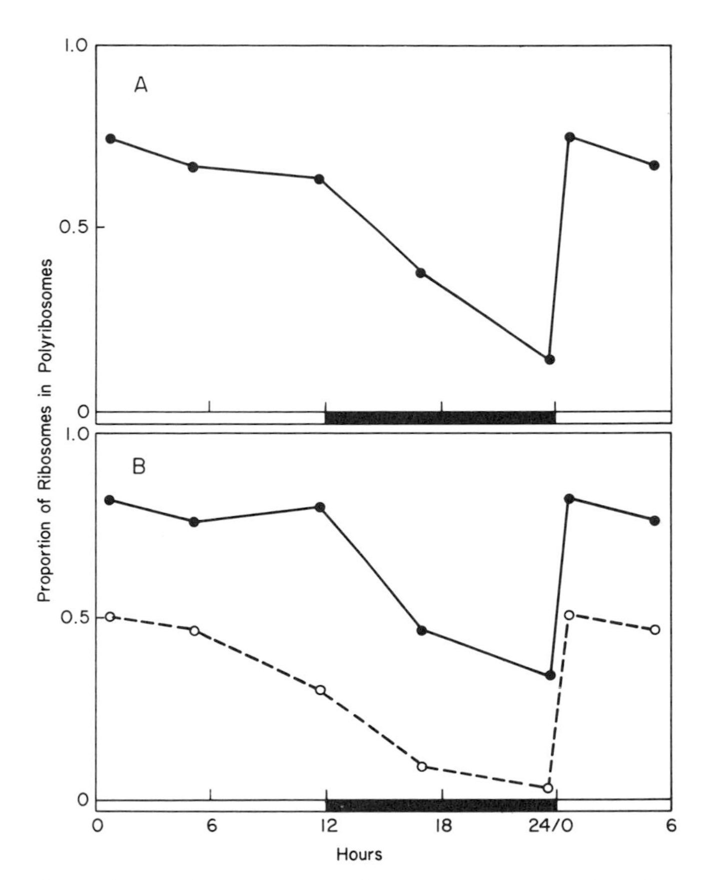

FIG. 2. Proportion of ribosomes in polyribosomes at different cell cycle stages. Total optical density (A_{260}) in monoribosomal (subunit) and polyribosomal fractions was determined from polyribosome profiles obtained from synchronous cells at different cell cycle stages. RNA from pooled monoribosome and polyribosome fractions was analyzed on polyacrylamide gels to separate chloroplast ribosomal RNA (16 S and 23 S) from cytoplasmic (18 S and 25 S). (A) The proportion of ribosomes in polyribosomes for both types of ribosomes. (B) The proportions for cytoplasmic ribosomes (●——●) and chloroplastic ribosomes (O---O). Cell division occurred as in Fig. 1 at about 16 hours or 4 hours into the dark period. [From Baumgartel and Howell (27).]

polyribosomes (see Fig. 2) at a time when r_e (see Fig. 1C) and the number of ribosomes per cell were increasing (33). At the beginning of the dark period (Fig. 3), r_i sharply fell and reached its lowest value halfway during the dark period. The drop in r_i occurred as r_e reached its greatest level, and both these factors, declining r_i and increasing r_e, led to a massive disassembly of polyribosomes.

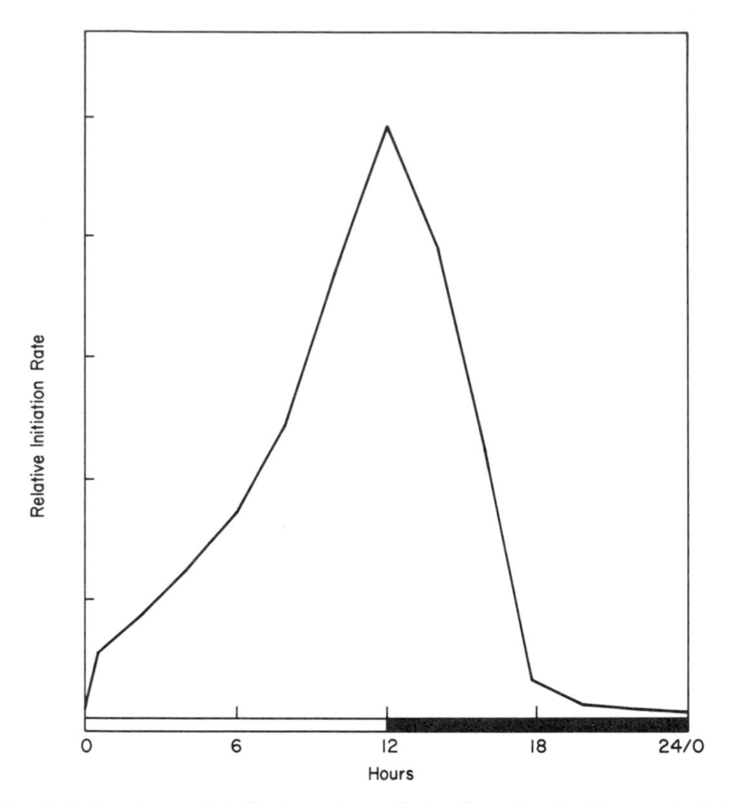

FIG. 3. Relative changes in initiation rate, r_i, during the cell cycle. R_i was calculated from the data in Figs. 1 and 2 and according to Eq. (2). See text for explanation. [From Baumgartel and Howell (27).]

These data permit us to calculate changes in the overall rate of protein synthesis, r_s (number of amino acids incorporated into peptide per cell per unit time), during the cell cycle. R_s is the product of the number of growing nascent chains per cell (or P, the number of translating ribosomes per cell) times their rate of growth, r_e, or

$$r_s = Pr_e \tag{3}$$

Substituting P in Eq. (2) into Eq. (3), we find that

$$r_s = mr_i \tag{4}$$

Thus, r_s varies with r_i, since we have assumed that m (the average size of translated mRNAs) is relatively constant during the cell cycle. Therefore, Fig. 3 is a plot of changes in the overall rate of protein synthesis, r_s, during the cell cycle, as well as changes in r_i. It is interesting to note that changes in r_s are quite different from changes in the rate of amino acid incorporation during the cell cycle (26, 27). Changes in amino acid incorporation (^{3}H-arginine) are only about five-to tenfold, and the pattern of incorporation is different than that illustrated in Fig. 3.

These experiments have shown us that the component rates of protein synthesis during the cell cycle in *C. reinhardi* are in a constant state of flux. We did not find, as might be expected from the previously described studies, that there was one rate of protein synthesis at mitosis and another rate established thereafter. Furthermore, the relationship between changes in the rate of protein synthesis and the cell cycle are not patently clear. Changes in r_i and r_e seem to correspond more closely to illumination conditions than to events of the cell cycle. At the time of cell division (about 4 hours in the dark), r_e reached its highest levels, but r_i and r_s were rapidly dropping. Whether the fall off in r_i and r_s relates to the onset of divisional activity is unclear. Instead, the decline in protein synthesis seems to be keyed to the beginning of the dark period (Fig. 3). The resumption of translation activity following division is certainly a light effect. The light period began at least 6 hours following cell division, and, at that time, there was a sudden reassembly of polyribosomes. It is possible that the reformation of polyribosomes following division might be delayed by the timing of the illumination cycle program, such that if synchronous cells were immediately shifted to light following division they might rapidly reform polyribosomes. These experiments, however, have not been done.

It is important to point out that both r_i and r_e fluctuated during the cell cycle but that changes in r_i were far greater than in r_e. It was particularly interesting as well to find out how the cell capitalizes on its ribosome population during the light period to achieve maximum efficiency of protein synthesis at a time when cell growth is the greatest. During the light period, r_i kept pace with both increasing r_e and the synthesis of new ribosomes to maintain a high proportion of ribosomes in polyribosomes.

B. Control of the Labeling of Single Polypeptide Species in the Cell Cycle

At any stage in the cell cycle, the overall rate of protein synthesis is a summation of the individual rates of synthesis of single polypeptide species. Changes in the overall rate of protein synthesis during the cell cycle are primarily a reflection of the rate changes in the synthesis of the major or most abundant polypeptide species. It is of interest to determine the rates of synthesis of many different polypeptide species with respect to the rates of overall protein synthesis. At issue is whether the syntheses of single polypeptide species are controlled completely by changes in the overall rate of protein synthesis.

We have looked into this question by examining, during the cell cycle, the rates of $^{35}SO_4^{3-}$ and [^{3}H]arginine labeling of polypeptides resolved by one-dimensional SDS–polyacrylamide gel electrophoresis. (Note: In this section we will be discussing rates of labeling, not rates of synthesis, since rate of synthesis measurements, as discussed previously, have not been carried out for individual polypeptides.) Figure 4 shows an autoradiograph of a polyacrylamide gel in which the relative rates of labeling (band density) of soluble and membrane polypeptides at different cell cycle stages can be ascertained. Cells at different cell cycle stages were pulse labeled for 25 minutes with $^{35}SO_4^{3-}$, and solubilized protein samples containing the same amount of radioactivity were loaded in separate gel channels.

Nearly 100 membrane and soluble polypeptides, representing the major or most abundant cell proteins, could be resolved on these gels (Fig. 4). Since the band densities of most polypeptides did not change significantly during the cell cycle, it can be concluded that most polypeptides were labeled at a constant relative rate (relative to all other polypeptide in the cell) during the cell cycle. Other polypeptides, however, showed considerable changes in relative labeling rates. Some notable examples are the membrane (M) polypeptides, such as M33, M31, M30, and M25, and the soluble (S) polypeptides such as S140, S84, and S39. These polypeptides and the others pointed out in Fig. 4, a total of about 20 out of 100, showed four- to tenfold changes in relative labeling rates during the cell cycle. (Only one polypeptide, S55, which is pointed out here, the large subunit of ribulosebisphosphate carboxylase, is synthesized by the chloroplast.)

The cell cycle map in Fig. 5 depicts the periods of peak labeling for those polypeptides which show labeling rate fluctuations in the cell cycle. It is interesting to note that the periods of relative maximal labeling for these polypeptides fall broadly into the mid-light period and the beginning of the dark period. It is during these intervals that changes in initiation rates (r_i) for total protein synthesis are the greatest.

At first inspection, the labeling of these polypeptides may appear to be a simple response to changes in illumination conditions. That may be partly correct. We have found (26) that many of the polypeptides shown in Fig. 5, which were labeled during the light period in synchronous cells, were labeled only in the light in asynchronous cells (designated by the symbol ○ in Fig. 5). Likewise, several polypeptides, which were labeled in the dark period in synchronous cells, were prominently labeled when light-grown asynchronous cells were shifted to the dark (designated by the symbol ●). There are exceptions such as S45 and S39 (Fig. 5), which fortunately spoil this simple explanation. Furthermore, it is not enough to state that illumina-

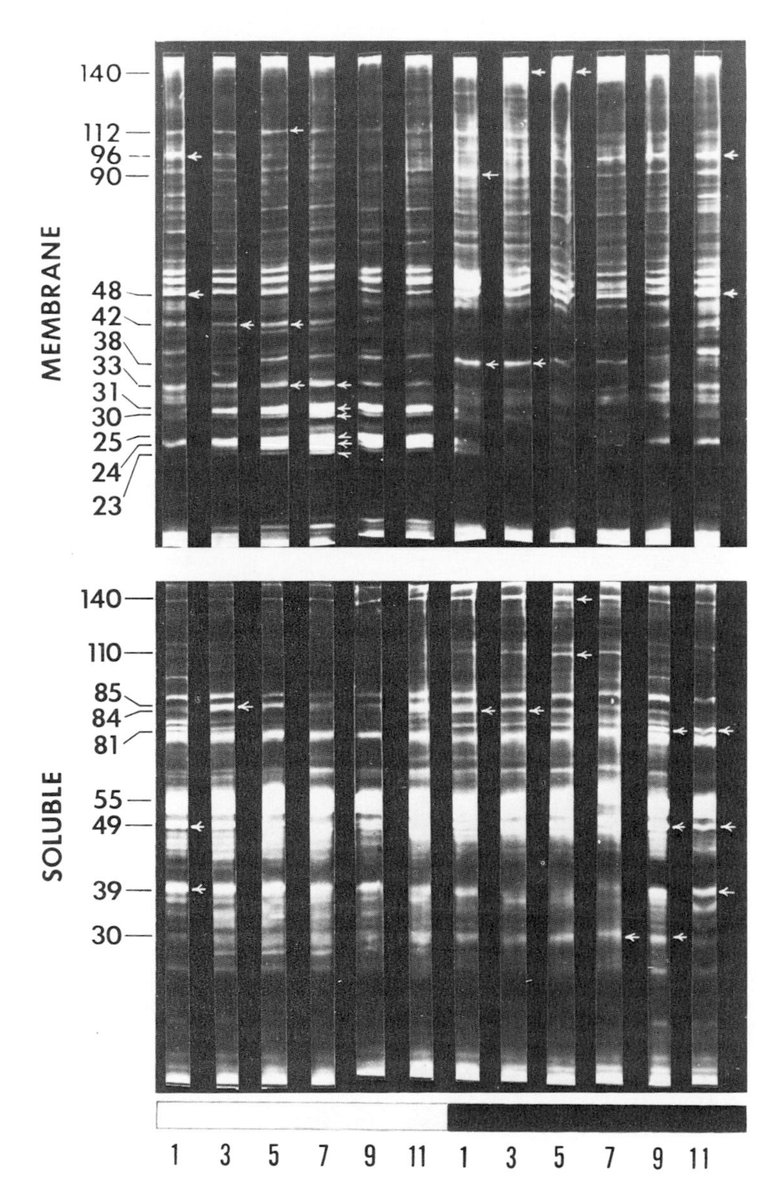

FIG. 4. SDS–polyacrylamide gel autoradiograph of membrane (P40) and soluble (S100) polypeptides pulse labeled with $^{35}SO_4^{3-}$ during the cell cycle. Cells synchronized with 12-hour light (open bar)–12-hour dark (dark bar) illumination cycle were pulse labeled for 25 minutes with $^{35}SO_4^{3-}$ at 2-hour intervals throughout the cell cycle (times of sampling are shown at the bottom). Samples containing the same amount of radioactivity were loaded into separate channels and subjected to electrophoresis on 11% polyacrylamide gels. Molecular weight designations ($\times$ 10^{-3}) are related to electrophoretic mobility based on migration of standards. Cell division occurred at about 6-7 hours into the dark period. Arrows show stages of maximum labeling (maximum band density) for polypeptides that show significant labeling variations in the cell cycle. [From Howell *et al.* (26).]

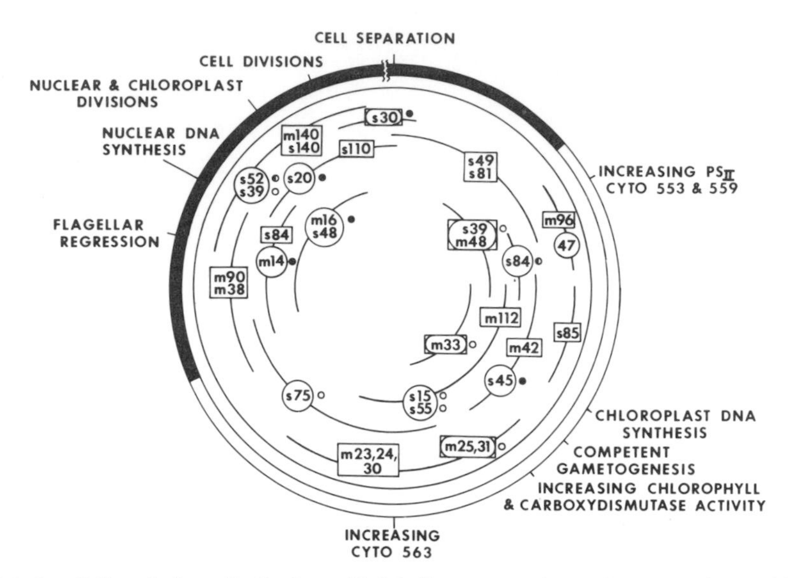

FIG. 5. Cell cycle "map" of polypeptide labeling patterns in synchronous cultures. Shown are the authors' interpretation of the period of maximum relative labeling of selected membrane (M) and soluble (S) polypeptides from experiments such as that shown in Fig. 4. ^{35}S-labeling experiments are designated as a rectangle, as S84; [^{3}H]arginine labeling is designated by a circle, as S75. The numbers refer to bands in Fig. 4. For comparison, cell cycle events (from Howell *et al.*, 25) are shown with relationship to 12-hour light and 12-hour dark illumination cycle. Small dot following the polypeptide designations indicate whether polypeptide is labeled in the light (○) or dark (●) in asynchronous light grown cells. Polypeptides labeled in the dark–light transition after asynchronous cells had been grown for 8 hours in the dark are indicated as ◐. [From Howell *et al.* (26).]

tion shifts are totally responsible for the periodic polypeptide labeling in these cells because the periods of peak labeling are not generally found at the light transitions, but are scattered at various times thereafter. One polypeptide, S84, did show marked light-transition-labeling behavior. It was labeled in synchronous cells at the dark–light transition and in synchronous cells upon shifting from dark to light conditions (designated by the symbol ◐ in Fig. 5). So, for the bulk of the polypeptides which are periodically labeled, it appears that cell cycle dynamics and not simply light transitions govern their labeling behavior.

These observations on the labeling of single polypeptide species raise the question asked in the beginning of this discussion as to whether the periodic synthesis of single polypeptides during the cell cycle is controlled by translational means. For most of the major cellular polypeptides, which were labeled at the same relative rate at different cell cycle stages, their labeling pattern may be regulated directly by the changing efficiency of the translational machinery during the cell cycle. For the polypeptides that showed periodic labeling patterns, they, too, may be regulated by overall rate changes in protein synthesis. There are two observations with regard to this issue that are worth bearing in mind. First, the periodic labeling of a polypeptide is really a change in the rate of labeling of that polypeptide, and is not an on-or-off response. The greatest changes in labeling rate for periodically labeled polypeptides were on the order of tenfold—not very different from the changes in the overall labeling rates. Transcriptional regulatory mechanisms might be expected to yield greater changes in the rate of labeling of some individual polypeptides, since transcriptional mechanisms are generally thought to be more specific and their effects more complete than translational mechanisms. Second, the labeling peak periods for most periodically labeled polypeptides occurred at times when overall initiation rates change most rapidly (compare Figs. 3 and 5). These observations raise the possibility that the periodic labeling of some polypeptides may be affected by, but might not necessarily follow, the changes in the overall rate of initiation. What is suggested is that the vast changes in the rate of initiation can lead to changes during the cell cycle in the competition of mRNA's to form initiation complexes. Such competition among mRNA's has been recently described by Lodish (34), who showed that changes in initiation rates in the reticulocyte system can alter the relative rates of α- and β-globin synthesis. Whether such competition regulates the translation of some mRNA's during the cell cycle is suggested, but this is, of course, completely speculative.

Our purpose has been to suggest the possible importance of translational mechanisms in controlling protein synthesis during the cell cycle. It would be naive to expect that the synthesis of every polypeptide could be controlled by strictly translational means. It has been our attempt here, however, to underscore the regulatory role that the rather dramatic changes in translation may play in the cell cycle of *C. reinhardi*.

ACKNOWLEDGMENTS

This work was supported by Grant GB-30237 from the National Science Foundtion.

REFERENCES

1. Weissbach, H., and Ochoa, S., *Annu. Rev. Biochem.* **45,** 191 (1976).
2. Lodish, H. F., *Annu. Rev. Biochem.* **45,** 39 (1976).
3. Mitchison, J. M., "The Biology of the Cell Cycle," p. 159. Cambridge Univ. Press, London and New York, 1971.

4. Halvorson, H. O., Carter, B. L. A., and Tauro, P., *Adv. Microb. Physiol.* **6,** 47 (1970).
5. Kedes, L. H., Gross, P. R., Cognetti, G., and Hunter, A. L., *J. Mol. Biol.* **45,** 337 (1969).
6. Stein, G. S., Spelsberg, T. C., and Kleinsmith, L. J., *Science* **183,** 817 (1974).
7. Stein, J. L., Thrall, C. L., Park, W. D., Mans, R. J., and Stein, G. S., *Science* **189,** 557 (1975).
8. Klevecz, R. R., *J. Cell Biol.* **43,** 207 (1969).
9. Petersen, N. S., McLaughlin, C. S., and Neirlich, D. O., *Nature* (*London*) **260,** 70 (1976).
10. Fraser, R. S. S., *Eur. J. Biochem.* **60,** 447 (1975).
11. Greenberg, J. R., *J. Cell Biol.* **64,** 269 (1975).
12. Mitchison, J. M., and Creanor, J., *J. Cell Sci.* **5,** 373 (1969).
13. Creanor, J., May, J. W., and Mitchison, J. M., *Eur. J. Biochem.* **60,** 487 (1975).
14. Prescott, D. M., and Bender, M. A., *Exp. Cell Res.* **26,** 260 (1962).
15. Fan, H., and Penman, S., *J. Mol. Biol.* **50,** 655 (1970).
16. Salb, J. M., and Marcus, P. I., *Proc. Natl. Acad. Sci. U.S.A.* **54,** 1353 (1965).
17. Johnson, T. C., and Holland, J. J., *J. Cell Biol.* **27,** 565 (1965).
18. Steward, D. L., Shaeffer, J. R., and Humpfrey, R. M., *Science* **161,** 791 (1968).
19. Hodge, L. D., Robbins, E., and Scharff, M. D., *J. Cell Biol.* **40,** 497 (1969).
20. Lodish, H. F., and Jacobsen, M., *J. Biol. Chem.* **247,** 3622 (1972).
21. Palmiter, R. D., *J. Biol. Chem.* **247,** 6770 (1972).
22. Fraser, R. S. S., and Carter, B. L. A., *J. Mol. Biol.* **104,** 223 (1976).
23. Brewer, E. N., *Biochim. Biophys. Acta* **277,** 223 (1976).
24. Mittermayer, C., Braun, R., Chayka, T. G., and Rusch, H. P., *Nature* (*London*) **210,** 1133 (1966).
25. Howell, S. H., Blaschko, W. J., and Drew, C. M., *J. Cell Biol.* **67,** 126 (1975).
26. Howell, S. H., Posakony, J. W., and Hill, K. R., *J. Cell Biol.* **72,** 223 (1977).
27. Baumgartel, D. M., and Howell, S. H., *Biochemistry* **16,** 3182 (1977).
28. Inger, M. D., and Tobey, R. A., *Biochemistry* **11,** 269 (1962).
29. Petersen, N. F., and McLaughlin, C. F., *J. Mol. Biol.* **81,** 33 (1973).
30. Kuff, E. L., and Roberts, N. E., *J. Mol. Biol.* **26,** 211 (1967).
31. Baumgartel, D. M., and Howell, S. H., *Biochim. Biophys. Acta* **454,** 338 (1976).
32. Vassart, G., Dumont, J. E., and Cantraine, F. R. L., *Biochim. Biophys. Acta* **247,** 471 (1971).
33. Cattolico, R. A., Senner, J. W., and Jones, R. F., *Arch. Biochem. Biophys.* **156,** 58 (1973).
34. Lodish, H. F., *Nature* (*London*) **351,** 385 (1974).

10

Regulation of Glutamate Dehydrogenase Induction and Turnover during the Cell Cycle of the Eukaryote *Chlorella*

Daniel W. Israel, Richard M. Gronostajski, Anthony T. Yeung, and Robert R. Schmidt

I. INTRODUCTION

A primary research goal in this laboratory is to elucidate the biochemical mechanisms that regulate gene expression during the eukaryotic cell cycle. Recent research activity has been centered on the problem of how eukaryotic microorganisms with long cell cycles and high rates of protein synthesis, such as *Chlorella,* regulate the expression of inducible genes and the levels of inducible enzymes during the cell cycle. For these studies, we have selected an inducible NADPH-specific glutamate dehydrogenase (GDH), which is absent in cells cultured in nitrate medium, but which can be induced by ammonium (1). Because this inducible GDH is not directly involved in carbohydrate metabolism, we consider it to be a better choice for cell cycle studies than inducible enzymes involved in hydrolysis of disaccharides (e.g., maltose, sucrose, etc.) or are otherwise involved directly in carbohydrate metabolism (2–6). Catabolite repression (7) might complicate interpretation of cell cycle studies with these latter enzymes.

The experimental evidence (1,8) presented in this chapter indicates that expression of the gene of the inducible GDH in *Chlorella* is regulated differently in cells cultured in the absence of, rather than in the continuous presence of, inducer. In the continuous presence of inducer, negative regulatory systems become operative and lower the rate of induced enzyme accumulation and delay the timing of expression of newly replicated inducible genes. The delay in timing of gene expression is proposed to be caused by oscillations in concentration of a repressing metabolite that accumulates in cells in the presence of inducer. Negative regulation by oscillatory repression was originally proposed (9–11) to explain the timing of expression of genes of biosynthetic rather than inducible enzymes during the cell cycle.

Whereas bacteria lower the levels of inducible enzymes primarily by dilution through cell divisions after removal of inducer from fully induced cells, we have revealed (8) that *Chlorella* has a process which very rapidly lowers the level of the NADPH-specific GDH during deinduction in the absence of cell division. In nondividing cells of higher eukaryotes, hormone-inducible enzymes also show rapid decreases in level after hormone withdrawal (12). The low frequency of cell division in eukaryotes might have led to the evolution of additional negative regulatory mechanisms not required to regulate the levels of inducible enzymes in prokaryotes.

II. INITIAL INDUCTION KINETICS OF THE NADPH-SPECIFIC GDH DURING THE CELL CYCLE

By use of polyacrylamide gel electrophoresis and a tetrazolium assay system, Talley *et al.* (1) showed that the NADPH-specific GDH was inducible throughout the cell cycle. Actinomycin D- and cycloheximide-inhibition experiments indicated that enzyme induction was dependent upon both RNA and protein synthesis. Moreover, induced enzyme accumulation was shown to be proportional to the rate of total protein accumulation in cells with a constant gene dosage.

If the structural gene of this enzyme is continuously available for transcription and if the rates of posttranscriptional processes do not vary enough in the cell cycle to obscure the maximal rate of gene transcription, a reasonable prediction is that the initial rate of induced enzyme accumulation should be proportional to the gene dosage during the cell cycle. Moreover, if genes can be translated shortly after replication, and if the time lag between transcription and translation is relatively short, the initial rate of enzyme induction would be predicted to increase within the S phase.

To test these predictions, cells were periodically harvested from a parent synchronous culture, growing in the absence of inducer, and then challenged to synthesize the NADPH-specific GDH. At each stage in the cell cycle analyzed, the enzyme was observed to accumulate in a linear manner for at least 80 minutes, following a 35-minute induction lag. The rate between 35 and 60 minutes was taken as a measure of the initial rate of induction at each stage of development. When the initial rates of induction were compared to the pattern of DNA accumulation in a synchronous culture in which each cell was dividing into four daughter cells, the initial rate of induction remained constant throughout the G_1 phase and then abruptly increased fourfold early within the S phase of the cell cycle (1). Thus, these findings are consistent with the aforementioned predictions.

In one respect, these findings with *Chlorella* make "biological sense," but from another viewpoint they are puzzling. It seems reasonable for a free-living eukaryotic cell to have evolved a mechanism by which genes, coding for enzymes involved in utilization of exogenous substrates, are inducible throughout the cell cycle. This type of regulation would give the cell the competitive advantage of utilizing exogenous substrates encountered at any time during its cell cycle. The puzzling feature, however, is that the gene of the NADPH-specific GDH is inducible during the period of nuclear division (13,14). Is it possible that certain inducible genes in the nucleus can be transcribed when chromosomes (or portions thereof) are in a condensed state? Or, is this inducible gene located in an organelle other than the nucleus? Alternatively, the induction might not be occurring at the transcriptional level during the period of nuclear division, as indicated by the studies with actinomycin D.

III. PATTERN OF ACCUMULATION OF THE NADPH-SPECIFIC GDH IN SYNCHRONOUS CELLS CULTURED IN THE CONTINUOUS PRESENCE OF INDUCER

Two general experimental approaches can be used to study the availability of inducible genes for transcription during the cell cycle. The first approach consists of culturing cells in the absence of inducer and then challenging them to synthesize the inducible enzyme at frequent intervals during the cell cycle. The second approach involves continuously challenging preinduced synchronous cells to synthesize an inducible enzyme by culturing them in the continuous presence of inducer for the entire cell cycle. In this latter approach, the type of cell cycle pattern of enzyme accumulation obtained and the relationship of this pattern to the S phase may give an insight into the availability of genes for transcription during the cell cycle. As discussed earlier (15), to interpret these enzyme-accumulation patterns, it is essential to know the *in vivo* stability of the inducible enzyme and its mRNA.

When preinduced synchronous *Chlorella* cells were cultured at a growth rate of 26% per hour in the continuous presence of ammonium (8), the NADPH-specific GDH increased in a linear pattern with positive rate changes preceding the onset of the periods of DNA accumulation by 1.5 hours in two consecutive cell cycles (Fig. 1). A linear pattern of induced enzyme accumulation is consistent with continuous synthesis of a stable enzyme and its unstable mRNA, or continuous synthesis of unstable enzyme and its stable mRNA. A third possibility, not discussed earlier by Schmidt (15), is continuous synthesis of stable enzyme with periodic synthesis of its stable mRNA.

Because Talley *et al.* (1) had observed the intial rate of induction of this enzyme to increase in early S phase, it was anticipated that the positive rate changes in fully induced enzyme accumulation would also occur within the period of DNA replication. Surprisingly, the positive rate changes actually occurred prior to the major periods of DNA replication. Therefore, they possibly reflect either delayed expression of genes replicated in the previous cell cycle or expression of genes replicated within the same cell cycle but prior to the major S phase. In an attempt to distinguish between these two possibilities, 2′-deoxyadenosine, an inhibitor of *in vivo* DNA replication (16–18), was added (2 m*M*) to fully induced cells approximately 2.5 hours prior to the first positive rate change in enzyme accumulation. This compound is converted *in vivo* to 2′-dATP, which is a potent feedback inhibitor of ribonucleotide reductase activity (19).

Although DNA synthesis was inhibited significantly in the first cell cycle, induced enzyme accumulation was not inhibited until the second rate change in the second cell cycle (Fig. 2). A close correlation between the magnitude of the second-rate change in enzyme accumulation and the fold increase in DNA was observed in both the control and 2′-deoxyadenosine-treated cultures, i.e., 3.6 versus 3.6 and 1.6 versus 1.8, respectively. When a higher concentration (2.75 m*M*) of the inhibitor was employed, similar correlations were observed between the magnitude of inhibition of DNA syn-

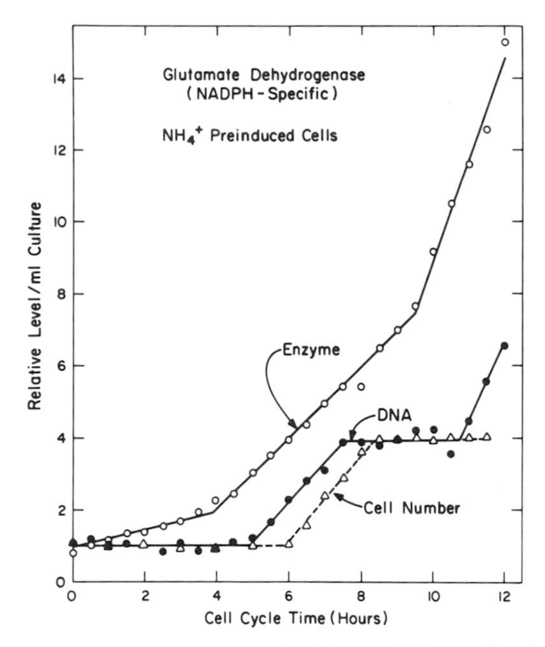

FIG. 1. Patterns of accumulation of an inducible NADPH-specific glutamate dehydrogenase and of total cellular DNA during the cell cycle of synchronous *Chlorella* cells growing at a rate of 26% per hour in the continuous presence of inducer (ammonium). The cells were precultured in ammonium medium during the light–dark synchronization procedure prior to selection of synchronous daughter cells by isopycnic centrifugation at the end of the third dark period. During the cell cycle study, the culture was maintained in a Plexiglas chamber in continuous light (15,900 lux), and its turbidity was held essentially constant by hourly dilutions with ammonium medium. The initial values per milliliter of culture for NADPH-specific glutamate dehydrogenase activity (○), DNA (●), and cell number (△) were 63 units, 17.3 μg, and 211 $\times 10^6$ cells, respectively. [From Israel *et al.* (8).]

thesis and inhibition of induced enzyme accumulation in the second but not in the first cell cycle (8). Although DNA synthesis and the pattern of enzyme accumulation was altered by the inhibitor, the patterns of increase in total cellular protein and in culture turbidity were essentially identical in the control and 2′-deoxyadenosine-treated cultures throughout the 12-hour experiments (8).

Since Israel *et al.* (8) showed that the inhibition of the rate change in the second cell cycle cannot be explained by the intracellular accumulation of 2′-deoxyadenosine or its phosphorylated derivatives with resultant inhibition of GDH activity per se, the proportional relationship between the magnitude of inhibition of DNA synthesis and the rate change in the enzyme accumulation in the second cell cycle is consistent with the inference that the structural gene of this inducible enzyme is expressed in the cell cycle after its replication in cells cultured at a growth rate of 26% per hour in the continuous presence of inducer.

Although the magnitude of the positive rate changes in enzyme accumulation appears to be related to gene replication in the previous cell cycle, the timing of these rate changes is influenced by factors other than gene replication. When the growth rate of the cells was lowered from 26% to 13% per hour by decreasing the intensity of incident light upon the culture, the positive rate change occurred within instead of outside the S phase (Fig. 3). The sensitivity of the timing of the rate changes to the growth rate of the cells supports, in part, the oscillatory repression model of inducible gene regulation discussed later in this chapter.

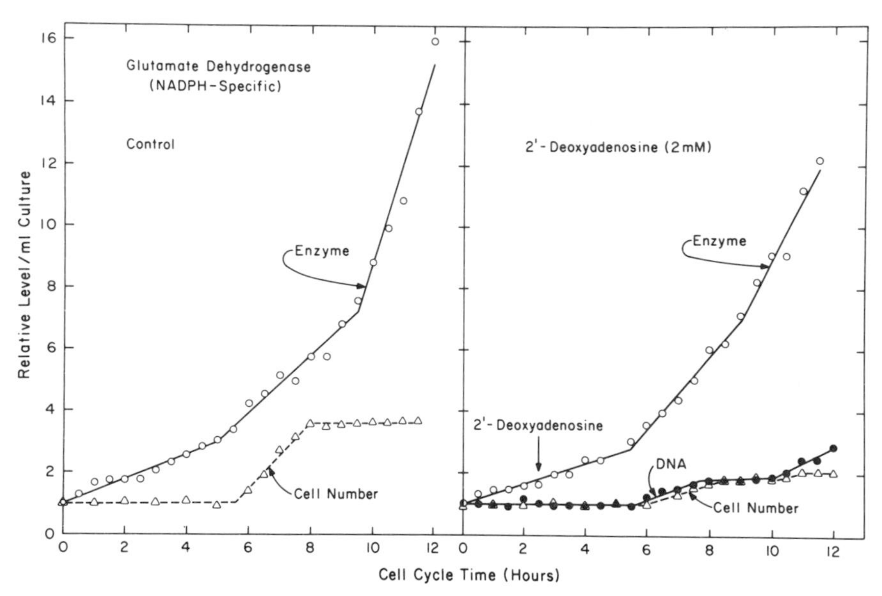

FIG. 2. Effect of 2 m*M* 2′-deoxyadenosine on the patterns of accumulation of an inducible NADPH-specific glutamate dehydrogenase and total cellular DNA during the cell cycle of synchronous *Chlorella* cells cultured in the continuous presence of inducer (ammonium). The synchronization procedure and culture conditions were the same as those described in Fig. 1. Equal volumes of the same suspension of synchronous daughter cells were placed into two identical Plexiglas chambers, and cell growth was initiated simultaneously in both cultures. After 2.5 hours of synchronous growth, 2′-deoxyadenosine was added in culture medium to one culture to bring its final concentration to 2 m*M*. Subsequent hourly dilutions of the treated culture were made with medium containing 2 m*M* 2′-deoxyadenosine. The control culture was diluted with medium without the inhibitor. The initial values per milliliter of culture of NADPH-specific glutamate dehydrogenase activity (○), DNA (●), and cell number (△) were 200 units, 14 μg, and 214×10^6, respectively. [From Israel *et al.* (8).]

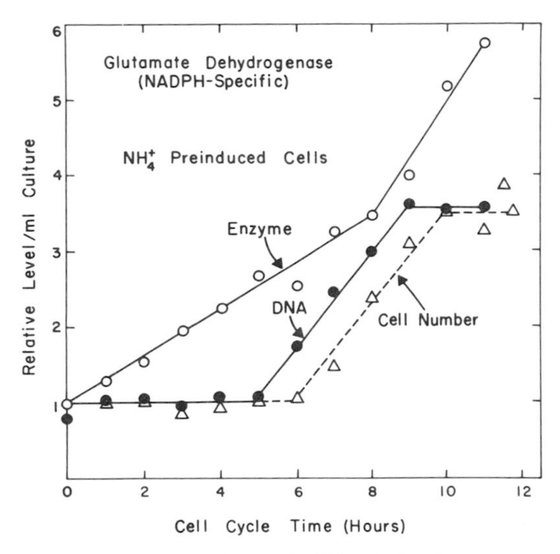

FIG. 3. Patterns of accumulation of an inducible NADPH-specific glutamate dehydrogenase and of total cellular DNA during the cell cycle of synchronous *Chlorella* cells growing at a rate of 13% per hour in the continuous presence of inducer (ammonium). The culture and synchronization procedures were the same as described in Fig. 1, except the light intensity was decreased to 6600 lux. The initial values per milliliter of culture for NADPH-specific glutamate dehydrogenase activity (○), DNA (●), and cell number (△) were 75 units, 13.5 μg, and 154×10^6 cells, respectively. [From Israel *et al.* (8).]

IV. *IN VIVO* STABILITY OF THE NADPH-SPECIFIC GDH DURING THE CELL CYCLE

Since the observed linear patterns of induced accumulation of the NADPH-specific GDH can be generated by either a stable or unstable enzyme, depending upon the stability of its mRNA, an attempt was made to ascertain the *in vivo* stability characteristics of the enzyme by blocking its synthesis with cycloheximide. Israel *et al.* (8) observed that cycloheximide inhibited accumulation of both GDH isozymes, and that these enzymes were stable *in vivo* for at least 6 hours in the absence of cellular protein synthesis (Fig. 4A,B).

Since the inducible NADPH-specific GDH appeared to be stable *in vivo*, it seemed possible that an estimate of the maximum half-life of its mRNA *in vivo* could be obtained. For example, from the rate of deceleration in rate of accumulation of stable β-galactosidase, after deinduction by addition of an anti-inducer, Jacquet and Kepes (20) obtained an estimate of the *in vivo* half-life of the β-galactosidase mRNA in *Escherichia coli*. When *Chlorella*

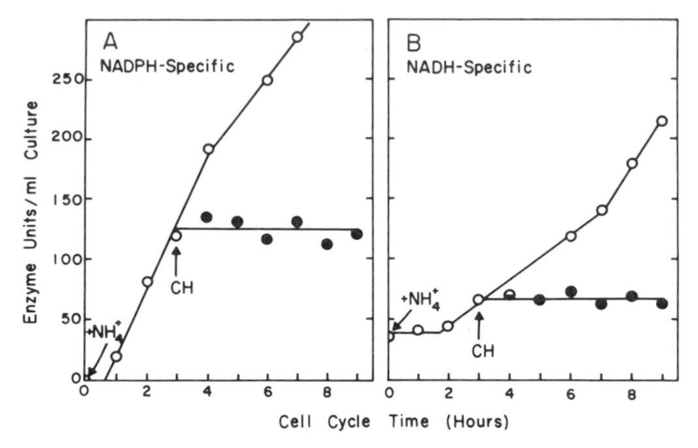

FIG. 4. *In vivo* stability characteristics of an ammonium-inducible NADPH-specific and a NADH-specific glutamate dehydrogenase in synchronized cells of *Chlorella* in ammonium medium after inhibition of total protein accumulation by cycloheximide. The cells were precultured and light–dark synchronized in nitrate medium. During cell cycle study, the culture of synchronized cells was maintained in a Plexiglas chamber in continuous light (6600 lux), and its turbidity was held essentially constant by hourly dilutions with ammonium medium. After 3 hours of synchronous growth, the culture was divided equally between two chambers. At this time, cycloheximide was added in culture medium to one culture to bring its final concentration to 0.089 m*M* (i.e., 25 μg/ml). Subsequent hourly dilutions of the treated cultures were made with medium containing 0.089 m*M* cycloheximide. The control culture was diluted with medium without the inhibitor. (A) and (B) NADPH-specific or NADH-specific glutamate dehydrogenase activity, respectively, measured in cells in the presence (●) and absence (○) of cycloheximide. The initial cell number per milliliter of culture was 140×10^6 cells. [From Israel *et al.* (8).]

cells were deinduced by removal of ammonium from the culture, the activity of the NADPH-specific GDH decayed very rapidly instead of continuing to accumulate at a decelerating rate (Fig. 5A). The other GDH isozyme continued to accumulate without change in rate (Fig. 5B), indicating that the NADPH-specific isozyme is not converted to the NADH-specific isozyme during the deinduction period.

Since nitrate was added to provide the cells with an alternate nitrogen source after removal of ammonium from the culture, the possibility existed that addition of nitrate caused the loss in activity of the NADPH-specific GDH. However, addition of nitrate to an ammonium-induced culture had only a slight effect on the rate of accumulation of this enzyme (Fig. 6). Moreover, after removal of ammonium, the activity of the enzyme decayed at the same rate ($t_{1/2}$ = 10 minutes) in cultures which were nitrogen deficient or contained nitrate (8). Thus, it is the removal of ammonium and not the addition of nitrate that triggers the loss in activity of the inducible GDH. The addition of cycloheximide at the time of inducer removal prevented the loss in enzyme activity during the deinduction period (Fig. 6).

The loss in activity of the NADPH-specific GDH could be due to (a) reversible inactivation [e.g., covalent modification (21,22)], or (b) irreversible inactivation or proteolytic degradation of the enzyme. To distinguish between these two possibilities, a rescue experiment with ammonium was performed. At approximately the midpoint of the deinduction period, ammonium was reintroduced with or without cycloheximide to a deinduced culture. The timing and kinetics of reinitiation of subsequent induced enzyme accumulation are consistent with irreversible inactivation or degradation of the enzyme during the deinduction period (Fig. 7). If the enzyme had been reversibly inactivated, its activity would have been anticipated to increase at a much faster rate than in an uninduced culture to which ammonium was added. This experiment does not rule out the possibility, however, that the enzyme was first reversibly inactivated and then rapidly degraded.

Dialysis of homogenates of cells, harvested prior to and after the deinduction period, did not reveal the presence of small molecule stabilizers or inhibitors of enzyme activity. Moreover, the activity of the enzyme was stable *in vitro* at 38.5°C in crude or dialyzed cell homogenates or cell

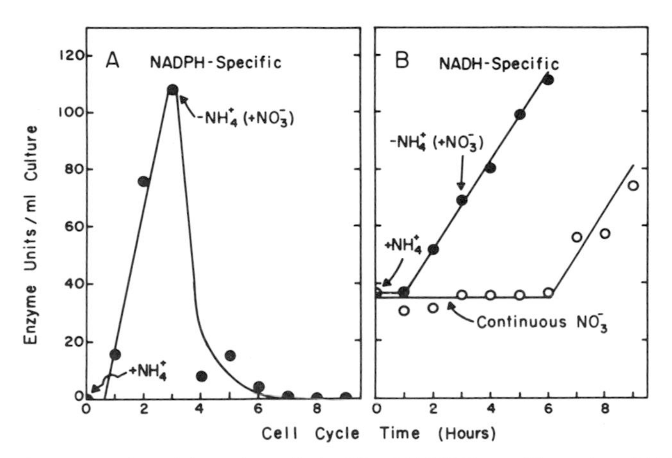

FIG. 5. Patterns of activity of an inducible NADPH-specific and a NADH-specific glutamate dehydrogenase in synchronized cells of *Chlorella* before and after removal of ammonium from the culture medium. The synchronized cells, and preculture and other culture conditions were essentially the same as those described in Fig. 4. After 3 hours of synchronous growth in ammonium medium, the cells were harvested by centrifugation and resuspended in an equal volume of medium without ammonium but with an equimolar concentration of nitrate. A separate culture (O) was maintained in nitrate medium throughout the induction and deinduction experiment. (A) and (B) NADPH-specific or NADH-specific glutamate dehydrogenase activity, respectively. The initial cell number per milliliter of culture was 133 × 10^6 cells. [From Israel *et al.* (8).]

extracts. Thus, the activity of the system responsible for the *in vivo* loss in enzyme activity does not appear to be active after cell disruption.

The inhibition of protein synthesis by cycloheximide has been reported to lower the rate of turnover of certain proteins in animal cells (23). Because cycloheximide blocked the *in vivo* decay in activity of the NADPH-specific GDH during the deinduction period (Fig. 6), the possibility exists that the inhibitor also might have prevented turnover of this enzyme in cells in the continuous presence of inducer (Fig. 4A). Thus, the *in vivo* stability characteristics of this enzyme in the presence and absence of inducer will have to be estimated by a more direct procedure, such as radioimmunoprecipitation (24).

During deinduction *in vivo*, cycloheximide might act directly or indirectly to prevent the loss in activity of the NADPH-specific GDH. It might act indirectly by (a) blocking the synthesis of enzymes responsible for inactivation or degradation of the inducible GDH or (b) inhibiting protein synthesis, thereby preventing utilization of small molecules (e.g., amino acids) which stabilize the inducible GDH against the action of existing inactivating or degradative enzymes. Alternatively, cycloheximide could act directly by (a) inhibiting the activity of existing inactivating or degradative enzymes, or (b) binding to the inducible GDH and maintaining it in a conformation which is not susceptible to inactivation or degradation. This inhibitor has been

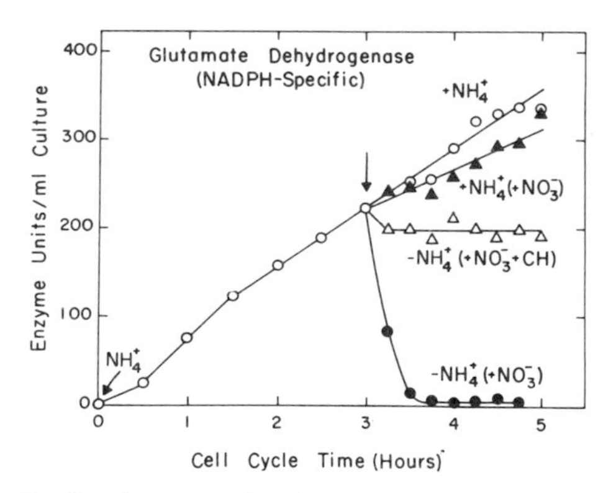

FIG. 6. Factors affecting the *in vivo* loss in activity of an ammonium-inducible NADPH-specific glutamate dehydrogenase in synchronized cells of *Chlorella* deinduced by removal of ammonium from the culture medium. The synchronized cells, and preculture and other culture conditions were essentially the same as those described in Fig. 4. After 3 hours of synchronous growth in ammonium medium, the culture was divided into four equal volumes, and the cells in each were harvested by centrifugation and resuspended in equal volumes of medium containing ammonium (○), nitrate (●), nitrate plus 0.089 m*M* cycloheximide (▲). Ammonium and nitrate were added in equimolar concentrations. The initial cell number per milliliter of culture was 172×10^6 cells. [From Israel *et al.* (8).]

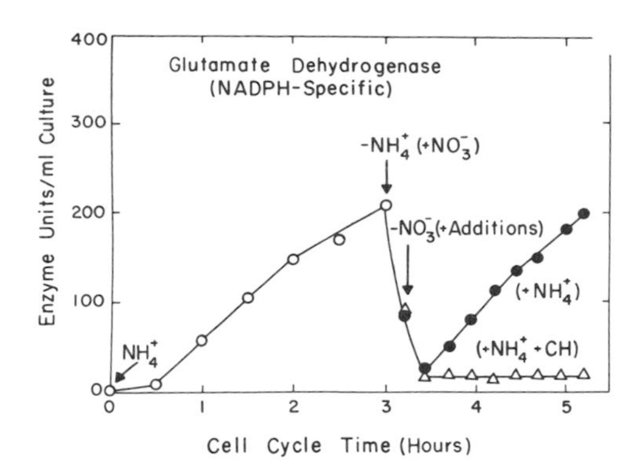

FIG. 7. Kinetics of reinduction of an ammonium inducible NADPH-specific glutamate dehydrogenase in synchronous cells of *Chlorella* following a short period of deinduction by removal of ammonium from the culture medium. The synchronized cells and preculture and other culture conditions were essentially the same as those described in Fig. 4. After 3 hours of synchronous growth in ammonium medium (○), the culture was harvested by centrifugation and the cells resuspended in the same volume of nitrate medium. After 12 minutes of deinduction under growth conditions, the culture was divided into two equal volumes, centrifuged, and the cells resuspended in equal volumes of ammonium medium (●) or ammonium plus 0.089 m*M* cycloheximide medium (△). The initital cell number and the activity of the isozyme per milliliter of culture were 176×10^6 cells and 209 units, respectively. [From Israel *et al.* (8).]

reported (25, 26) to affect the activity of a number of enzymes (e.g., alcohol and lactate dehydrogenases, ribosomal RNA polymerase) and of other cellular processes (27) in a number of different organisms.

V. INDUCTION KINETICS AND EVIDENCE FOR A REPRESSING METABOLITE

The negative rate change in accumulation of the NADPH-specific GDH,observed 1.5–4 hours into the induction period in all experiments (Figs. 4A, 6, and 7), is consistent with the accumulation of a repressing metabolite, which is synthesized directly or indirectly from ammonium. This negative rate change also was seen after ammonium was reintroduced to a deinduced culture (Fig. 7). Because induction measurements of Talley *et al.* (1) were made for only 80 minutes after addition of inducer to previously uninduced cells, they did not observe this negative rate change. The timing of the negative rate change also appears to be dependent upon the growth rate of the cells.

If the NADPH-specific GDH is stable in cells in the continuous presence of inducer (Fig. 4A), the negative rate change reflects a reduced rate of synthesis of this inducible enzyme. The negative rate change probably signals the onset of a regulatory system that operates to prevent the overproduction of an inducible enzyme in cells growing in the continuous presence of inducer.

VI. MODEL FOR DELAYED EXPRESSION OF THE STRUCTURAL GENE OF THE NADPH-SPECIFIC GDH IN CELLS CULTURED IN THE CONTINUOUS PRESENCE OF INDUCER

085

We propose that expression of the structural gene of the NADPH-specific GDH is regulated by an inducer that is required for gene expression, and a metabolite that accumulates in the presence of inducer and represses gene expression. The following statements formulate the basis of a model which is considered to be tentative and is presented with the idea of stimulating further experimentation (Fig. 8). In cells cultured in the absence of inducer, the repressing metabolite is absent or at low levels. Upon addition of inducer to the medium of previously uninduced cells, the initial rate of induced enzyme accumulation is proportional to the existing gene dosage at any time during the cell cycle. Under these conditions, inducible genes are expressed shortly after replication. In cells cultured in the continuous presence of inducer, the repressing metabolite, which is synthesized either directly or indirectly from the intracellular inducer, accumulates to levels sufficient to repress gene expression. Because the repressing metabolite is part of a feedback loop system in which this metabolite inhibits the absorption of inducer and/or its own synthesis, the intracellular level of the repressing metabolite oscillates. The timing and magnitude of these oscillations are influenced by the rate of utilization of both the repressing metabolite and the inducer for synthesis of other cellular constituents. These rates of utilization are, in turn, influenced by the metabolic state and/or the growth rate of the cells. In this feedback loop system, gene expression varies inversely with the peak concentrations of repressing metabolite. Thus, if the concentration of repressing metabolite is high during the S phase, the full expression of newly replicated structural genes of the inducible enzyme will be delayed until a later period during which the repressing metabolite decreases to its minimal concentration.

Although the model is general in nature, it gives rise to certain predictions that are experimentally testable. The following predictions are typical examples. First, the step increase in the initial rate of induction (i.e., maximal rate of enzyme accumulation after addition of inducer to previously uninduced cells) of the NADPH-specific GDH would be predicted to be inhibited within the S phase of the same cell cycle in which DNA

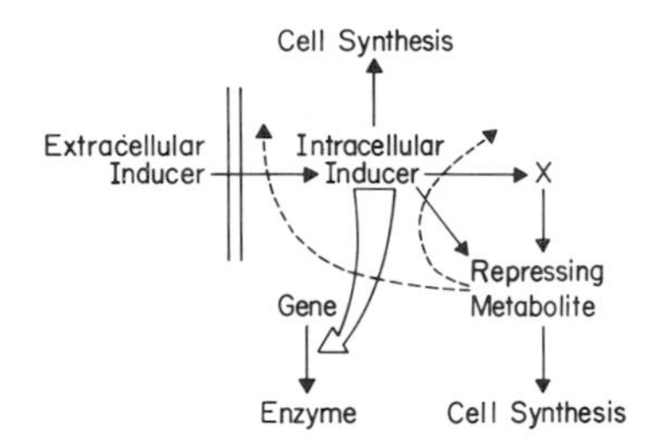

FIG. 8. Model for regulation of expression of the structural gene of an ammonium-inducible NADPH-specific glutamate dehydrogenase in *Chlorella*. Expression of the structural gene is induced (open arrow) by an intracellular inducer. A repressing metabolite, which is synthesized either directly or indirectly from the inducer, accumulates in the presence of inducer and can repress (---) gene expression. In cells cultured in the absence of extracellular inducer, the repressing metabolite is normally absent or at low levels, so that the initial rate of induced enzyme synthesis is proportional to the gene dosage existing at any time during the cell cycle. Because the repressing metabolite is part of a feedback loop system in which it inhibits (---) its own synthesis, the intracellular level of the repressing metabolite oscillates in cells cultured in the continuous presence of extracellular inducer. The timing and magnitude of the oscillations are influenced by the growth rate of the cells, which, in turn, affects the rate of synthesis of the repressing metabolite and its rate of utilization for synthesis of other cellular constituents. Since gene expression varies inversely with the peak concentrations of repressing metabolite, the expression of newly replicated genes can be displaced beyond the S phase for a length of time dependent upon the periodicity of the oscillations in level of repressing metabolite. The extracellular and intracellular inducers could be identical or different compounds.

synthesis is inhibited. Second, when the inducer concentration range is varied below that which supports maximal rates of induced enzyme accumulation in cells growing in the continuous presence of inducer, the timing of the rate changes in linear pattern (i.e., timing of delay in gene expression) would be predicted to vary in cells growing at the same rate. In this latter case, the demand on the inducer or its metabolites for growth would be predicted to lead to decreased levels of the repressing metabolite and/or a change in timing of its internal fluctuations. In the studies described in the present chapter, the timing of the rate changes was varied by changing the growth rate of the cells in the continuous presence of saturating exogenous levels of inducer. Third, at a specific exogenous concentration of inducer and cell growth rate, the timing of the rate changes in linear enzyme accumulation should be altered by supplying compounds (e.g., glutamate, glutamine, etc.) to the cells which are synthesized *in vivo* into or from the repressing metabolite (or intracellular inducer). The addition of these compounds would presumably perturb the timing of the endogenous oscillations of repressing metabolite. Fourth, it should be possible to isolate mutants that cannot accumulate high levels of repressing metabolite, either because their rate of synthesis is greatly reduced or their utilization or breakdown is enhanced. If the repressing metabolite levels are reduced in these mutants, such that these levels approach the level in uninduced cells, the timing of the rate changes should occur within the same S phase in which the genes of the inducible enzyme are replicated. In other words, the rate changes in linear enzyme accumulation in mutant cells growing in the continuous presence of inducer should occur near the time of the midpoint for the step increase in enzyme potential for wild-type cells.

The step at which the repressing metabolite limits gene expression in *Chlorella* could occur at any one of a number of biochemical levels (i.e., transcription, processing of precursor mRNA, translation, enzyme turnover, etc.). To locate this step, the turnover rate of the inducible enzyme and its mRNA must be determined along with the amount of specific translatable mRNA in different cellular fractions during the cell cycle. Moreover, the mRNA of the inducible enzyme should be purified (28–30) and cDNA made from it (31) so that this cDNA can be used in hybridization studies (32) to locate and to quantify mRNA sequences that might exist in a nontranslatable state in different cellular fractions during the cell cycle under different inducing conditions. The cDNA also can be used to locate the cellular site (i.e., nucleus, chloroplast, mitochondria) of the structural gene of the NADPH-specific GDH.

VII. COMPARISON OF INITIAL INDUCTION KINETICS AND FULLY INDUCED ENZYME ACCUMULATION IN OTHER EUKARYOTIC MICROORGANISMS

Halvorson and co-workers (2, 33–35) also have made a study in a eukaryotic microorganism in which the initial kinetics of induction and the pattern of induced enzyme accumulation was compared in previously uninduced and fully induced synchronous cells, respectively. They observed continuous inducibility of enzymes in synchronous cells of *Saccharomyces* previously cultured in the absence of inducer. However, all inducible enzymes examined exhibited step instead of linear patterns in synchronous cells cultured in the continuous presence of inducer.

We propose that these seemingly paradoxical results can be unified by the assumption that, in the continuous presence of inducer, oscillations in the level of a repressing metabolite result in step patterns of an inducible enzyme from either periodic synthesis of unstable mRNA, with resultant periodic synthesis of stable enzyme, or periodic synthesis of stable mRNA with continuous synthesis of unstable enzyme. Although certain experimental findings (34) tend to argue against this proposal, they do not eliminate oscillatory repression as a possible mechanism for regulating inducible enzyme levels in synchronous cells of *Saccharomyces* cultured in the continuous presence of inducer.

VIII. DELAYED GENE EXPRESSION OBSERVED DURING MEASUREMENT OF INITIAL INDUCTION KINETICS DURING THE CELL CYCLE

From our model (Fig. 8), one would not anticipate some of the long delays observed between the timing of gene replication and gene expression in measurement of the initial rates, i.e., enzyme potential (9, 15), of induction (or derepression) in previously uninduced (or nonderepressed) cells of several different eukaryotic microorganisms (3–6, 36–38). Without being inconsistent with the model, however, there are a number of possible reasons which can be given to explain these delays.

First, after addition of inducer to previously uninduced cells, if the rate of synthesis of the repressing metabolite is very rapid relative to the rate of induced enzyme synthesis, the repressing metabolite could reach repressive levels and inhibit induced enzyme accumulation during the period of measurement of enzyme potential. Cell cycle variations in the rate of synthesis of the repressing metabolite relative to that of the inducible enzyme could cause an apparent delay in the increase in enzyme potential. In fact, some of the longest delays observed (3, 37) between the period of DNA replication and the increase in enzyme potential have been in studies in which the length of the period of induced enzyme accumulation was long, relative to the length of the cell cycle. In enzyme potential measurements on repressible rather than inducible enzyme systems, cell cycle variations in the rate of utilization of the intracellular corepressor, after removal of the extracellular precursor to the corepressor, could also result in apparent delays in gene expression (3, 37, 38).

Second, in some organisms, the repressing metabolite might be a normal intermediate in cellular metabolism and might be synthesized by an alternate pathway in cells growing in the absence of inducer. In uninduced cells growing rapidly in the absence of inducer under constant culture conditions, the rate of utilization of the repressing metabolite might be high relative to its rate of synthesis so that it does not accumulate to repressive levels. However, if this alternate pathway were under negative feedback control by the repressing metabolite (or one of its metabolites), dramatic changes in the growth rate of the cells could result in accumulation of the repressing metabolite and subsequent oscillations in its level for a period of time even in uninduced cells. Consistent with this possibility is the dramatic change in the timing of the increase in enzyme potential in *Schizosaccharomyces pombe* synchronized by different procedures (3, 16, 17). Moreover, the delay in the increase in enzyme potential for isocitrate lyase in two species of *Chlorella* (4, 36) and the inability to induce the synthesis of this enzyme during certain periods of the cell cycle of a third species of *Chlorella* (5) might have been due to the accumulation of a repressing metabolite when the cells were shifted from photosynthetic to non-photosynthetic conditions for the induction measurements. Alternatively, since isocitrate lyase accumulation cannot be induced in any of these species of *Chlorella* cultured under photosynthetic conditions, the repressing metabolite might accumulate in cells in the light in the absence of inducer and might be utilized in the dark during those periods in the cell cycle in which synthesis of the enzyme is inducible. Thus, cell cycle variations in the rate of dark utilization of the repressing metabolite could account for the observed delay in the rise in the enzyme potential or in the inability to induce synthesis of this enzyme at certain periods of the cell cycle.

Third, the previous discussion has dealt primarily with apparent delays in gene expression seen in enzyme potential measurements. However, similar delays have been seen for repressible enzyme systems in which the basal rates of synthesis of repressible enzymes were measured in cells cultured in the presence of high levels of specific corepressors or their precursors. Under these conditions, rate changes in linear accumulation of different repressible enzymes were observed to occur after instead of within the DNA replication period (3). Again, these delays might be related to the operation of a negative feedback system which regulates the uptake and/or conversion of extracellular corepressor (or precursor) to active intracellular corepressor, so that, even in the excess of extracellular corepressor, the active intracellular level oscillates and displaces the timing relationship between gene replication and expression.

ACKNOWLEDGMENTS

The research discussed in this chapter has been supported by NIH Grant GM 19871 and Grant BMS 75-02287 from the National Science Foundation.

REFERENCES

1. Talley, D. J., White, L. H., and Schmidt, R. R., *J. Biol. Chem.* **247,** 7927 (1972).
2. Tauro, P., and Halvorson, H. O., *J. Bacteriol.* **92,** 652, (1966).

3. Mitchison, J. M., and Creanor, J., *J. Cell Sci.* **5,** 373 (1969).
4. Baechtel, F. S., Hopkins, H. A., and Schmidt, R. R., *Biochim. Biophys. Acta* **217,** 216 (1970).
5. McCullough, W., and John, P. C. L., *Biochim. Biophys. Acta* **269,** 287 (1972).
6. Woodward, J., and Merrett, J. M., *Eur. J. Biochem.* **55,** 555 (1975).
7. Perlman, R. L., and Pastan, I., *Curr. Top. Cell. Regul.* **5,** 117 (1971).
8. Israel, D. W., Gronostajski, R. M., Yeung, A. T., and Schmidt, R. R., *J. Bacteriol.* **130,** 793 (1977).
9. Kuempel, P. L., Masters, M., and Pardee, A. B., *Biochem. Biophys. Res. Commun.* **18,** 858 (1965).
10. Goodwin, B. C., *Nature* (*London*), **209,** 479 (1966).
11. Donachie, W. D., and Masters, M., *in* "The Cell Cycle: Gene-Enzyme Interactions (G. M. Padilla, G. L. Whitson, and I. L. Cameron, eds.), p. 37. Academic Press, New York, 1969.
12. Tomkins, G. M., Gelehrter, T. D., Granner, D., Martin, D., Samuels, H. H., and Thompson, E. B., *Science* **166,** 1474 (1969).
13. Curnutt, S. G., and Schmidt, R. R., *Exp. Cell Res.* **36,** 102 (1964).
14. Schmidt, R. R., *in* "Cell Synchrony: Studies in Biosynthetic Regulation" (I. L. Cameron and G. M. Padilla, eds.), p. 189. Academic Press, New York, 1966.
15. Schmidt, R. R., *in* "Cell Cycle Controls" (G. M. Padilla, I. L. Cameron, and A. Zimmerman, eds.), p. 201. Academic Press, New York, 1974.
16. Mitchison, J. M., and Creanor, J., *Exp. Cell Res.* **67,** 368 (1971).
17. Sissons, C. H., Mitchison, J. M., and Creanor, J., *Exp. Cell Res.* **82,** 63 (1973).
18. Wanka, F., *Exp. Cell Res.* **85,** 409 (1974).
19. Reichard, P., *Adv. Enzyme Regul.* **10,** 3 (1972).
20. Jacquet, M., and Kepes, A., *Biochem. Biophys. Res. Commun.* **36,** 84 (1969).
21. Schutt, H., and Holzer, H., *Eur. J. Biochem.* **26,** 68 (1972).
22. Segal, A., Brown, M. S., and Stadtman, E. R., *Arch. Biochem. Biophys.* **161,** 319 (1974).
23. Epstein, D., Elias-Bishko, S., and Hirsko, A., *Biochemistry* **14,** 5199 (1975).
24. Schimke, R. T., *in* "Methods in Enzymology" (B. W. O'Malley and J. G. Hardman, eds.), Vol. 40, p. 241. Academic Press, New York, 1975.
25. Latuasan, H. E., and Berends, W., *Rec. Trav. Chim. Pays-Bas* **77,** 416 (1958).
26. Horgen, P. A., and Griffin, D. H., *Proc. Nat. Acad. Sci. U.S.A.* **68,** 338 (1971).
27. Ross, C., *Plant Physiol.* **53,** 635 (1974).
28. Shapiro, D. J., Taylor, J. M., McKnight, G. S., Palacious, R., Gonzales, C., Kiely, M. L., and Schimke, R. T., *J. Biol. Chem.* **249,** 3665, (1974).
29. Shapiro, D. J., and Schimke, R. T., *J. Biol. Chem.* **250,** 1759 (1975).
30. Woo, S. L. C., Rosen, J. M., Liarakos, C. D., Choi, Y. C., Busch, H., Means, A. R., and O'Malley, B. W., *J. Biol. Chem.* **250,** 7027 (1975).
31. Monahan, J. J., Harris, S. E., Woo, S. L. C., Robberson, D. L., and O'Malley, B. W., *Biochemistry* **15,** 223 (1976).
32. McKnight, G. S., Pennequin, P., and Schimke, R. T., *J. Biol. Chem.* **250,** 8105 (1975).
33. Halvorson, H. O., Bock, R. M., Tauro, P., Epstein, R., and LaBerge, M., *in* "Cell Synchrony: Studies in Biosynthetic Regulation" (I. L. Cameron and G. M. Padilla, eds.), p. 102. Academic Press, New York, 1966.
34. Sebastian, J., Carter, B. L. A., and Halvorson, H. O., *Eur. J. Biochem.* **37,** 516 (1973).
35. Carter, B. L. A., and Halvorson, H. O., *Exp. Cell Res.* **76,** 152 (1973).
36. Aasberg, K. E., Lien, T., and Knutsen, G., *Physiol. Plant.* **31,** 245 (1974).
37. Mitchison, J. M., and Creanor, J., *Exp. Cell Res.* **69,** 244 (1971).
38. Mitchison, J. M., *Symp. Soc. Gen. Microbiol.* **23,** 189 (1973).

11

The Effect of Different Cell Cycles in Yeast on Expression of the Cytoplasmic *Petite* Mutation[a]

Byron F. Johnson, Allen P. James, Norman T. Gridgeman, C. V. Lusena, and Eng-Hong Lee

I. INTRODUCTION

Respiratory deficient (RD, see Table I for terminology) strains of yeast have been known for about 40 years (1–4). The classic studies of Ephrussi and his group (5, 6) indicated that the inheritance of the *petite colonie*

[a] N.R.C.C. No. 15832.

TABLE I

Terminology of Relevant Cytoplasmically Inherited Characteristics[a]

Terms		
Wild type	Mutant	Characteristics
Grande	*Petite*	Colonial phenotype
RS (respiratory sufficient)	RD (respiratory deficient)	Cellular phenotype
Rho$^+$ (ρ^+)	*Rho*$^-$ (ρ^-)	Mitochondrial genotype

[a] Examination of the biologic characteristics shows that the mutant terms are clearly nonsynonymous. Nevertheless, synonymy is often assumed, and, indeed, little ambiguity results, for a cell whose mitochondrial genotype is ρ^- has defective respiration and generates a clone whose colony is *petite*.

A variety of drug-resistance markers and other mitochondrial "point" mutations are now known, but these lie outside the terms of reference of this chapter.

mutation (5, Table I) is non-Mendelian, that the mitochondrion is the locus (6) of many of the physiological defects [lack of cytochromes a, a_3, b, and certain respiratory enzymes (7), lack of cristae (8), excess of cytochrome c (7)], and demonstrated the complexity of the mutational patterns among the progeny of single cells (5). Ephrussi's studies converted respiratory deficiency from a curiosity to a subject of broad genetic, biochemical, and cellular interest.

Many criteria are used (Table II) to establish that the inheritance of the character is extrachromosomal, or "cytoplasmic;" the strongest is the loss of the character at conjugation (Fig. 1a). Ephrussi's first crossings (5) of haploid cells carrying the *petite* mutation with normal haploids yielded diploids that were normal (Fig. 1a). This seemed to indicate simple dominance of wild type over *petite* (or ρ^+ over ρ^-, Table I), but subsequent

TABLE II

Criteria for Cytoplasmic (Extrachromosomal) Inheritance in Yeast[a]

I. Non-Mendelian segregation
   A. Neutral *petites*: loss of character
   B. Suppressive *petites*: variable loss of character
II. Heterokaryon test
   Phenotype variably associated with nuclear markers
III. The unstable state; high specific induction rates;
   (Acriflavine, EtBr, UV, high temperature, etc.)
IV. Detectable alteration of cytoplasm
   A. Ultrastructural changes of mitochondria
   B. Modification or loss of mitochondrial DNA

[a] Adapted from Jinks (47).

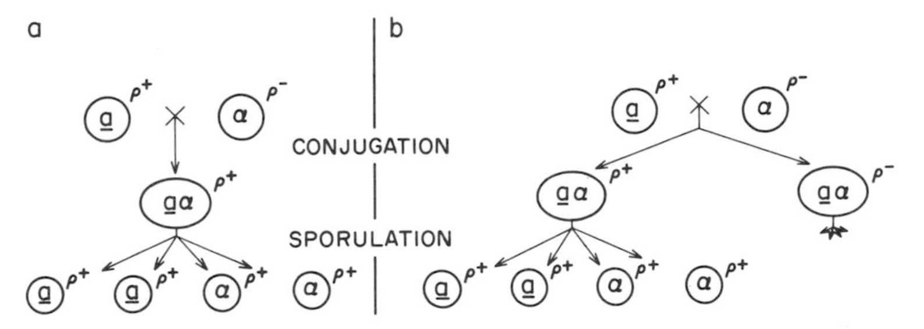

FIG. 1. Crosses between *grande* (ρ^+) and *petite* (ρ^-) haploids. Mating type genes (α and *a*) show normal Mendelian inheritance, segregating equally among the haploid spores. Note the contrasting inheritance of ρ factors in 1a, a nonsuppressive mating, and in 1b, a suppressive mating. The magnitude of suppressiveness is defined by the percentage of *petite* isolates among the diploids. The ρ^- haploid conjugating in 1a, thus, is defined as 0% S, or neutral. [Adapted from Ephrussi and Hottinguer (5).]

sporulation revealed that the mutant character had actually disappeared in the mating: all four meiotic products were normal.

Later, Ephrussi *et al.* (9) showed that certain ρ^- strains yielded more complex results. Thus (Fig. 1b), both ρ^+ and ρ^- diploids were produced by "suppressive" strains when mated with the wild type, but the ρ^+ diploids again yielded only ρ^+ meiotic products. Suppressive matings, thus, indicated that the inheritance of this character is indeed complex, but in no way is its mode of inheritance Mendelian. Since those early studies, other RD mutations with cytoplasmic inheritance (mit$^-$) and some with nuclear, or Mendelian, inheritance have been discovered, but they lie outside the frame of reference of this chapter, and further discussion relates to the cytoplasmic petite (ρ^-) mutants only.

Physiological and cytological studies of the *petite* isolates showed that they were defective for respiration (5, 6), and the Nadi reaction, hence, suggested that they were entirely comparable with the mutants discovered earlier (1–4). Many years of study were required to show (1) that yeast mitochondria usually contain DNA (10, 11), which inferentially provides a biochemical basis for the cytoplasmic inheritance pattern, and (2) that mitochondria from RD cells either contain no DNA—hence, ρ^0 (12, 13)—or drastically modified DNA (usually enriched in adenine and thymine; 14). Intensive biochemical study of both normal yeasts and the *petite* mutants continues in many laboratories, so that a welter of details about changes in mitDNA, etc., is now available. Surprisingly, these have helped only a little in our understanding of the complex pedigrees described very early by Ephrussi and his co-workers (5).

Successive buds may be separated from an "adult" or "mother" yeast cell by micromanipulation, and each can be allowed to produce its own individual colony. A series of colonies, a mitotic pedigree, can, thus, be obtained from a known sequence of "daughter" cells. Ephrussi performed (5) an extensive and imaginative series of pedigree experiments with acriflavine compounds, which induced the *petite* mutation much more frequently than chromosomal genes could be mutated. By placing budding cells on agar containing amounts of acriflavine small enough to allow budding to occur at a normal rate, but also adequate to induce the *petite* mutation, and by allowing cells to remain exposed to acriflavine for varying periods of time, he obtained very complex patterns of *petite* and *grande* colonies in his pedigrees. These were ascribed to an enigmatic unstable state generated in the adult cell by the drug. Further experiments demonstrated that under certain conditions, and at strain-specific doses of acriflavine, pedigrees could be obtained in which the original adult cell remained *grande*, but essentially all progeny were *petite* (15). Serving to complicate matters further was the observation that some drugs induce the adult to become *petite* also (16).

Interest in this unstable state was recently renewed by a persuasive suggestion (17–19) that the *in vitro* Qβ replication system (20) serves as a model for a putative population of mitochondrial genomes (mitgenomes) within a yeast cell. In this "competition" model, the rate of replication of each mitgenome is related to the primary structure of its mitDNA, with mitDNA of the mutant having a higher replication rate, a circumstance which would lead to a high rate of production of ρ^- mutants. The competition model was well received and stimulated much discussion (21–23).

In this chapter, we summarize the results of a pedigree analysis of spontaneous *petite* production (the unstable state) done intensively enough to allow statistical analysis (24). Comparison of the results with the different cell cycles found in pedigrees suggests an alternative model for expression of the cytoplasmic *petite* mutation. The model is based upon different regulations of mitDNA metabolism, rather than differences in primary structure of the mitDNA

II. INTENSIVE STUDY OF SPONTANEOUS *PETITE* PRODUCTION IN PEDIGREES

A. Experimental Rationale

If a yeast cell contains a population of mitgenomes, some normal (ρ^+) and having a normal rate of replication, and some defective (ρ^-) with a different rate of replication, then the character of the mitgenome population should change as a function of time. To examine changes in the character of such a population, one merely isolates by micromanipulation a budding yeast cell (a mother, M) under conditions which permit production of *petites*, and from that cell removes its buds (daughters, D's) in sequence. Each bud is assumed to contain a sample of the mitgenome population of the progenitor cell. Hence, the trend toward more or fewer *petite* daughters as a function of daughter number should reflect the trend toward a higher or lower ρ^- proportion among the mitgenomes in the maternal population.

We made the experiment intensive by (1), examining a large number of partial mitotic pedigrees (a final total of 138); (2) examining 10 budding cycles (10 daughters) of the isolated mothers [a mean maximum of about 24 budding cycles per mother can by analyzed (25)]; and (3) examining the first granddaughter produced by each of the 10 daughters.

Our test also differed from previous pedigree analyses by our decision to avoid chemical or physical inducers, of which there are many (26); rather, we used a strain that inately casts petite progeny at high frequency (24). By avoiding chemical or physical agents as inducers, we hoped to avoid interpretive complications which are inevitably consequent to the assumed modes of interaction between the cells and the inducing agents.

Many more or less reasonable assumptions are required for interpreting the results in detail. It is obvious that data in pedigree experiments may be complex and will bear on many aspects of cellularity and *petite* production—but detailed exposition of the data and assumptions is elsewhere (24), and here we restrict ouselves to generalities.

B. Experimental Results

Of the 138 partial pedigrees established, 71 were rejected because all colonies were *petite,* and 67 were mixed—*grande* and *petite*—allowing analysis of trends. The *petite* frequency among granddaughters was about two or three times that of the daughters, and the *petite* frequency among daughters was about four times that of the mothers (49/67 versus 18/67; see Fig. 2 for details). This asymmetry was statistically highly significant and clearly resembled differences shown earlier by Sherman (27), who used exposure to high temperature to induce *petite* mutations.

The trends of *petite* frequency as a function of daughter number are most easily revealed by comparison of the first half-pedigree with the second half. Such comparison shows (Fig. 2) a significant (at the 3% level of significance) decrease in *petite* frequency in the later generations. However, the extent of this trend is diminished by the overall high frequency of *petite* progeny in the pedgrees. Restriction of the anlaysis to pedigrees with

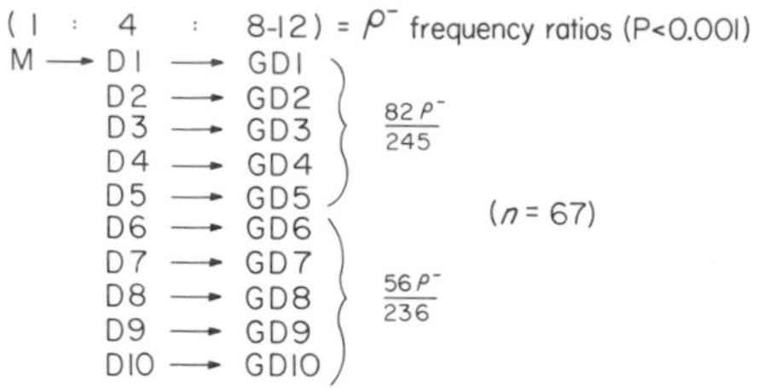

(1 : 4 : 8-12) = ρ^- frequency ratios (P<0.001)

M → D1 → GD1
D2 → GD2
D3 → GD3 $\frac{82\ \rho^-}{245}$
D4 → GD4
D5 → GD5
D6 → GD6 (n = 67)
D7 → GD7
D8 → GD8 $\frac{56\ \rho^-}{236}$
D9 → GD9
D10 → GD10

FIG. 2. A 10-cycle partial pedigree experiment with summarized results. M is the originally isolated budding adult (mother); D_1–D_{10} are the first 10 buds, daughters, in sequence; GD_1–GD_{10} is the sequence of the 10 first granddaughters (see text). A complete pedigree of 10 cycles equals 2^{10} cells (colonies), a logistically impossible micromanipulation experiment. [Adapted from James *et al.* (24).]

grande mother and four or fewer mutant progeny pairs (there were 36 of these) emphasizes the contrast in *petite* frequency between progeny $\text{pairs}_{1\text{-}5}$ (45/180, 25%) and progeny $\text{pairs}_{6\text{-}10}$ (16/172, 9%), the difference now being highly significant ($P \ll 0.001$).

In a few of our mixed pedigrees, the original cell (mother) became *petite* herself after generating some *grande* progeny. Their number, 18/67, was too small to permit analysis of maternal characteristics directly. However, comparison of their pedigrees with pedigrees produced by *grande* mothers shows that the probabilities of the production of a *petite* daughter by a *grande* mother are markedly different: only 68/481 (0.14) for *grande* mothers versus 38/76 (0.50) for *petite* mothers, If, as assumed, each bud samples the maternal mitgenome population, then this high ratio (0.50) means that these particular maternal populations of mitgenomes had high proportions of ρ^- to ρ^+, which should have been the case if the maternal cells themselves were becoming *petite*.

The results are inconsistent with the changes in the maternal population of mitgenomes predicted by the competition model (general *increase* in *petite* frequency with daughter number). If indeed the ρ^- mitgenomes are more efficient than the ρ^+ at replication, then some other factor obscures the predicted changes in the population.

In summary, we found a pronounced overall asymmetry in the pedigrees, with the original cell becoming, on the average, more *grande* in character and daughters of daughters having an enhanced capacity for being *petite*. It seems reasonable to suppose that this pattern of shifting instability is general to yeast strains and that its detection in this study is partly attributable to the fact that an inducing agent was not used as in previous investigations (5, 27). Protocols that utilize such agents tend to emphasize the chemical action on the cell rather than the cellular aspects per se.

III. CELL CYCLE REGULATION AND THE ASYMMETRIC PEDIGREES

A. A Segregation Model

When entertaining possible cellular bases for the asymmetry described above, one may be tempted to assume (28) that all cell cycles are identical, e.g., isolates 4, 1, and 1-1-1-1 (Fig. 3), all having been generated by identical budding cycles. This assumption easily leads to the conclusion that the asymmetry is being regulated by differential segregation; the adult cell retains elements that allow *grande* phenotypic expression, the bud receives elements that predicate *petite* phenotypic expression. Such elements would be the ρ factors, the normal (ρ^+) ones being retained, and the mutated (ρ^-) ones being transferred into the bud.

However, a more attractive alternative explanation of the asymmetry is available from the fact that budding yeast cell cycles are *not* identical. It has long been known that buds differ dramatically from their mothers in size (29; reviewed in 30–32, for examples), but, recently, quantitative data became available which allow comparisons within pedigrees (33).

B. A Differential Cell Cycle Model

1. Pedigrees and Morphometry

Budding yeast cell length and cell width data have been correlated with bud scar numbers (34), and, hence, with the progenal history of each cell. More recently, these data have been expressed in terms of cell volumes (33). When these data are plotted along the sides of a complete pedigree (Fig. 4), one sees immediately that the right side, the daughters of daughters, per-

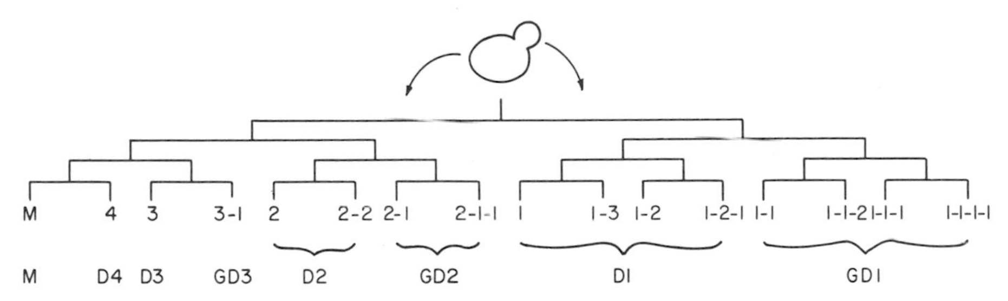

FIG. 3. A conventional, complete four-cycle pedigree with terminology of the partial pedigree infraposed below conventional terms, thus indicating which conventional clones are not separated in the partial pedigree. By convention, the adults are moved to the left, the buds to the right. [Adapted from James *et al.* (24).]

petually repeats the 0-scar cell volume, 206 μm^3, whereas the maternal side, the left side by convention, shows an increasing cell volume as a function of the numbers of daughters removed (or scars). So the asymmetry of the volume distribution resembles the asymmetry of the *petite* frequency distribution of the pedigree.

Unbalanced growth is implicit in this volume asymmetry, for the two products of cell division will, on the average, have different volumes, and most cell cycles will not involve a doubling of cell volume (35). After smoothing the data (from 33, 34), we establish a parameter for imbalance (I_i), thus: Let V_i be the volume of the mother cell at the i-th budding with $i = 0, 1, \ldots, 10$. We know V_0 and the mean volume increase, v, of the mother cell during a budding cycle. Let

$$I = \text{cell cycle volume increase}$$
$$= (V_i + v + V_o)/V_i$$

So if $V_o = 200$, and $v = 43$, we have, at the i-th budding cycle,

$$I_i = \frac{400 + 43i}{200 + 43(i - 1)}$$

yielding values of I_i ranging from 2.22 to 1.41 for $i = 1$ to 10, respectively (Table III). In considering this imbalance, we are less concerned with cell volumes than with the nature of mitDNA metabolism in such unbalanced cell cycles.

2. Regulation of MitDNA Replication in Unbalanced Cell Cycles

Is the amount of mitDNA so tightly coupled to the amount of nuclear DNA (nDNA) that its synthesis is regulated for precise doubling in every

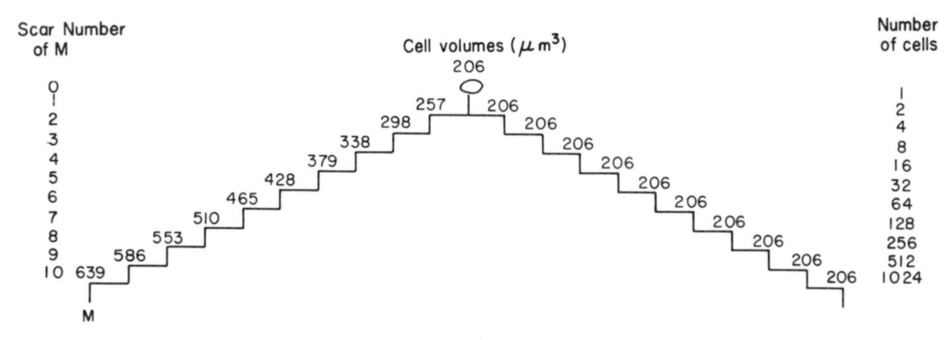

FIG. 4. Volumes of cells at extreme sides of a complete 10-cycle pedigree; 206 μm^3 is clearly the volume of unbudded daughter cells. (Data from 33, 34.)

TABLE III

Extent of Imbalance of Different Yeast Cell Cycles

i (Budding cycle[a])	I_i (Cell cycle volume increase)
1	2.22
2	2.00
3	1.85
4	1.74
5	1.65
6	1.59
7	1.53
8	1.49
9	1.45
10	1.41

[a] The 0-scar cell obviously is in its first (i = 1) budding cycle, the 9-scar in its tenth, etc. [Based upon data from Beran *et al.* (34) after smoothing—see text.]

cell cycle, or is the coupling to nDNA so loose that perhaps other cellular parameters are operative? Data exist (Table IV) which show that the amount of mitDNA is not tightly coupled to nDNA but, in fact, is correlated with cell volume, suggesting that the synthesis is feedback controlled by cytoplasmic demand for a mitochondrial product such as ATP. Because the volume increment is about the same for 0-scar mothers and 9-scar mothers in their budding cycles, the amounts of mitDNA synthesized will be comparable. However, from the asymmetry of relative cell volume changes (Fig. 4), their *relative extents of replication* of mitDNA differ by a factor of three: about 120% of the mitDNA is replicated during budding of 0-scar mothers, but only about 40% is replicated in the budding cycle of 9-scar mothers (see Table III). Our interpretation of how this asymmetry of mitDNA replication at the cellular level bears on the regulation of mitDNA metabolism is as follows.

3. Extrinsic Regulation of MitDNA Synthesis Rates

At the mitochondrial level, one might expect that, on the average, ρ^+ mitgenomes will be associated with respiratorily effective membranes surrounding competent matrices, and that ρ^- mitgenomes will be associated with respiratorily defective membranes and matrices. The relative access of mitgenomes to the deoxytrinucleotide substrates required for replication should parallel the matrical competence. This implies that the rate of mitDNA metabolism is regulated *extrinsically* by the associated matrix character, rather than regulated intrinsically by its own primary structure, and that ρ^+ mitgenomes may, on average, be found to replicate at a higher rate than the ρ^-.

TABLE IV

Comparison of Yeast Cell Volumes and Content of mitDNA

Condition	Δ Volume			Δ mitDNA			Comments
	μm^3	%	(Reference)	Amount and units	%	(Reference)	
Vegetative							
Synchronous culture adult → adult + bud	298 → 544	183	(33)	1.8 → 3.1 (μg/50 ml)	172	(12)	
Preconjugal							
α and *a* cells mixed	40 → 68	170	(40)				nDNA does *not* replicate in 180 minutes by this protocol
α-Factor-arrested *a* cells					200	(48)	150-minute observation
a-Factor-arrested *α* cells					300	(49)	180-minute observation
Conjugal							
Zygote formation	(2 × 40) → 240	300	(40)		300–400	(49)	
					150	(50)	nDNA doubles
Vegetative							
0-scar cells → large adult cells	55 → 135	245	(49)	0.20 → 0.35 (ratio mit:tot)	175	(49)	
Haploid:diploid comparison	13.8:26.0	188	(51)	38.8:68.5 (× 10^{-16} gm/cell)	177	(51)	Isogenic, save for ploidy

Likewise, matrical competence could also influence the rate of recombination among mitgenomes, for recombination is another enzymatic process whose rates will be regulated to some extent by the size of precursor pools. Recombination between homologous strands of mitDNA occurs extensively during synthesis (36–38), and recombination among neighboring mitgenomes continues to occur in diploid progeny over several cell cycles after conjugation (39); this strongly suggests a persistent link with mitDNA metabolism. The rates of recombination should be correlated with the rates of replication if they are subject to similar regulations.

4. The Model and the Asymmetric Pedigrees

We suggest that the population of mitgenomes in a yeast cell is modulated by (1) a volume control regulation which predicates a diminished extent of mitDNA replication as a function of increased cell age; (2) an extrinsic regulation of mitDNA biosynthesis, which favors the ρ^+ mitgenomes; and (3), extrinsically regulated recombination also favoring the ρ^+ mitgenomes.

Now if ρ^+ is superior to ρ^- both in replication and recombination, what we might call an "inhibition factor" operates as follows (see Table III for values of I):

If $I > 2$, replication and recombination are uninhibited, whereas, if $I < 2$, positive inhibition favors the more efficient mitgenomes, the net result being enhancement of the proportions of ρ^+ mitgenomes as i increases. Basically, we are suggesting that the resultant differential synthesis and recombination rates of ρ^+ and ρ^- mitDNA account, in part at least, for the dissimilar fates of mother and daughter. In particular, recombination involving ρ^- mitgenomes will be more frequent in the daughters than in the larger mother cells—independent of the initial ρ^+/ρ^- ratio. Recombination between the ρ^+ and ρ^- mitgenomes could have an important modulating effect; here lies the possibility of converting sense to nonsense, for *both* products might well be ρ^-. Seemingly, we are dealing with a mechanism for destroying the ρ^+ character of a mitochondrion, hence, for facilitation of expression of the ρ^- genotype through amplification of ρ^- mitgenomes, rather than by mutation in any normal sense of the word.

IV. OTHER CONSIDERATIONS

A. The Differential Cell Cycle Model and Suppressiveness

Suppressiveness is the production of *petites* during sexual cell cycles, a consequence of mating *grandes* with *petites*. It has been thought to be closely related to vegetative production of *petites* by the proponents of the competition model (18, 19). If this model is correct, then high mean suppressiveness should be found in clones having high mean mutational frequency. However, we have not found (B. F. Johnson and A. P. James, unpublished data) a correlation between the frequency of mutants in pedigrees and the mean suppressiveness of the isolates. Either the parameters for regulatory control systems differ or the controls themselves differ between sexual and vegetative cell cycles.

Cellular behavior during conjugation is obviously different (see Fig. 5) from that during mitosis. For instance, there is a tripling of the cell volume during mating of *grandes* (40). It is likely that similar volume changes occur during the mating cycle of *petite* cells, and it is even more likely that these changes will be the same when 10% S (suppressive) and 95% S cells are compared. Hence, the volume controls on mitgenome replication probably regulate identically in ρ^+ tester strains and in ρ^- strains regardless of their suppressiveness. If so, the extrinsic parameters that made the primary structure of the mitDNA trivial for regulation of vegetative generation of *petites* have also been canceled out. But surely, if there are differences in primary

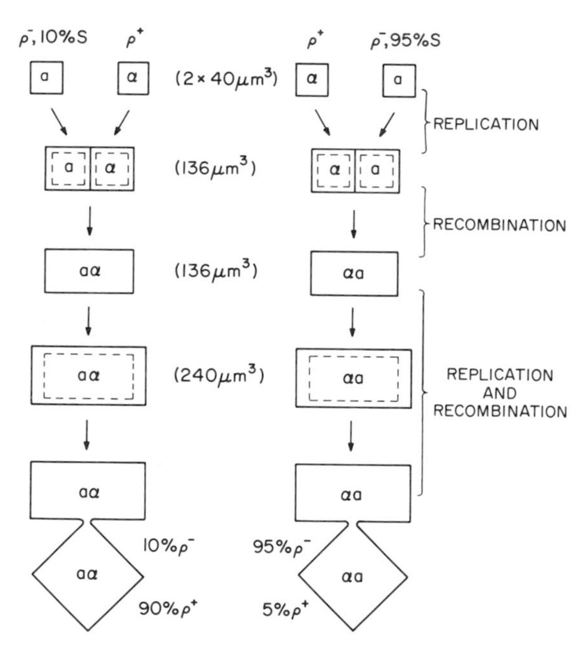

FIG. 5. Volume change models during two suppressive, isogenic matings. Volume changes of *grandes* during matings (40) are presumed to pertain to the *petites* as well. Areas indicate volumes.

structure between the two types of ρ^- mitDNA that relate to their different innate recombination potentials, the 95% S having the stronger potential, those differences may well become paramount when the imputed extrinsic controls over recombination neutralize one another. Thus, the volume controls and the extrinsic controls are presumed to regulate mitDNA metabolism throughout the mating process but not to bear on the expression of suppressiveness because they neutralize through identical regulations.

The *petite* progeny from suppressive matings yield no information, but the *grande* progeny are informative. Let us compare, for example, the fates of two new diploid cells, each of which has just enough ρ^+ character to qualify as *grande*, rather than *petite*. If the one *grande* derives from the 95% S parent and the other from the 10% S parent, will there be different rates *petite* production of their clones? The competition model suggests the affirmative, but the differential cell cycle model suggests the opposite. In fact, our own unpublished data (E.-H. Lee and B. F. Johnson), and, inferentially, the data of Forster and Kleese (41), indicate negligible differences in *petite* frequency among diploid *grande* clones, regardless of the suppressiveness of their ρ^- progenitor. The sexual production of *petites* through suppressive matings, thus, seems to be related to a different regulatory control parameter.

B. Cell Cycle Analysis and Mitochondrial Mutations

When analyzing cell cycles by size selection (42), it is usual to assume that there is only a single cell cycle typifying the population. However, the cell cycle of budding yeasts has been shown (30–32) to produce daughters which are immature in being unready to initiate their own first buds. These 0-scar daughters constitute one-half of the cellular population under ordinary cultural conditions and have a budless maturation phase. In contrast, old mothers are constantly budding, bud_{n+1} being initiated as soon as bud_n becomes a physiologically independent entity. Clearly, there is not a "typical cell cycle" but, rather, there are two typical cell cycles in a budding yeast population, a fact which may cause misinterpretation of data if cell cycles are analyzed by size selection alone.

Separation of small cells from large cells by gradient devices is intended to separate cells early in *the* cycle from cells at later stages in that cycle. But the small cells at the top of the gradient are really the unbudded portion of the 0-scar half of the total population. Thus, the gradient is actually separating the population into fractions that differ markedly from what they are usually assumed to be. For many experiments, this may make no difference. On the other hand, if induced *petite* mutation is under examination, then those top-of-the-gradient small cells are really from the daughter class that we have shown above to be more prone than their mothers to be *petite*. On such gradients, one is not seeing induction of *petite* mutations in the early phases of *the* cell cycle (43) but is seeing the maturing, *petite*-prone 0-scar cells which have yet to initiate their budding cycle. Furthermore, if point mutations of mitochondrial drug-resistance markers are established by physiological controls similar to those regulating the appearance of the *petite* mutations, then analysis of their mutational patterns with respect to the cell cycle by gradients is fraught with pitfalls. For instance, many of the 0-scar cells remain firmly attached to their mother through their maturation stage (or longer), while the mother goes on to initiate her next bud. Such a triad will be found near the bottom of the gradient; its colony will not be scored as *petite* if the mother retains her *grande* colonial phenotype, but the colony will be scored as drug resistant, even if only the 0-scar daughter was mutated. Unless one knows that 0-scar daughters have been physically removed with high efficiency from their mothers, such ambiguities becloud interpretation of selection cell cycle analyses for mitochondrial mutations.

V. GENERAL DISCUSSION AND CONCLUSIONS

We have examined a few of the many factors involved in regulating the production of cytoplasmic *petite* mutants in yeast. We have not considered the basic process of mutation of mitgenomes but have limited ourselves to a model which regulates by modulating the proportion of mutant genomes in the cell's mitochondrial population.

The differential cell cycle model for modulation of the mitgenome population seems to provide an adequate explanation for the asymmetric pedigrees (the unstable state) and the stability of *grande* diploids produced in suppressive matings. It seems likely that the differential cell cycle parameters for regulation may also influence conjugating yeast cells, but they do not have a detectable modulating role in the expression of suppressiveness, probably because their effects cancel out.

We emphasize again that the model is based upon conjecture as well as on fact. The size asymmetry in the pedigree is well established (33), as is the relationship between the amounts of mitDNA and yeast cell size (Table IV). Although there is no direct evidence for extrinsic regulation of mitDNA biosynthesis, such regulation would provide an explanation of the trend toward stability of the maternal cell in pedigrees. Nor is there evidence to support our speculation about extrinsic regulation of recombination between neighboring mitgenomes—paralleling the rate of synthesis—but again, it is plausible. Altogether, these assumptions do not seem onerous to us.

It is true that this model, an explanation of the production of mutant progeny, is based on the postulation of a series of probable physiological events that do not involve mutation at the molecular level. But then, it is not a model for molecular mutation. Rather, it is a model for regulating the biologic effect of premutated mitDNA by modulating the proportion of mutated to normal (or ρ^- to ρ^+) in a cellular population of mitgenomes, and, thus, it is a model for facilitating the expression of the *petite* phenotype. However, there is no question that, over and above this, genuine mitochondrial mutation at the molecular level does occur in the strain under discussion here, albeit at a very slow rate, for we have not been able to isolate a single genuinely stable clone from our strain. All of our "*grande*" colonies are actually mixtures of *grande* and *petite* cells, and, indeed, are visibly stratified (vertically sectored) like the *gi* colonies of Negrotti and Wilkie (44).

Doubtless, there are cases in which *petite* production (especially in response to some mutagens) involves physiological processes other than those detailed above, for instance, production of ρ^0 *petites*. Nevertheless, it seems likely, from the large number of agents known to induce the *petite* mutation in yeast (26), that many agents merely enhance the consequences of a sequence of activities for which the yeast cell is already programmed. Ephrussi's complex pedigrees, i.e., the unstable state, and Sherman's asymmetric pedigrees after heat induction may well be exaggerations of an innate process. *Petite* production by elevated pressure (45) may well fall into the same general class.

Studies in which maternal cells did not respond to the mutagen, but in which nearly all progeny were produced as *petite* (15), now seem best explained in terms of the mutagen helping to maintain the cellular proportion of ρ^+ mitgenomes/ρ^- mitgenomes at such a level that most daughters' mitochondrial populations would evolve toward ever higher frequencies of ρ^- mitgenomes, but the maternal population remaining with more or less constant proportions of ρ^+/ρ^-, thus, always retaining the *grande* phenotype in her maternal colony. On the other hand, those mutagens that cause "mother cells as well as buds to become *petite*" (16, 46) are perhaps more effective and mutate enough mitgenomes that the fraction of ρ^- mitgenomes in the maternal population gradually increases in the presence of the mutagen in spite of the tendency of volume regulations, etc., to favor her ρ^+ mitgenomes.

Of course, generation of a model for regulation is not an end in itself; it is, hopefully, only a beginning. It should stimulate other thoughts that may lead to its own obsolescence, and it should stimulate new experiments, either to expand its own data base, or to test reasonable predictions. We have under way a variety of tests of its predictions, but are impressed by the difficulty of expanding the data bases of the assumed extrinsic controls.

One reason for the failure of previous explanations, over a quarter of a century, is that all cell cycles have always been assumed to be identical. But this is contrary to our knowledge of continuous yeast cell growth, which has been known (29) for a quarter of a century! One could expand the model by involving the activities of enzymes known or reasonably assumed to act in a regulatory way. However. it seems premature to discuss regulation of a process by enzymic activities when we are ignorant of how those enzymes' syntheses, activations and degradations are themselves regulated.

We have attempted to show how different cell cycles of the budding yeast are involved in regulating some of the processes that facilitate expression of the cytoplasmic *petite* phenotype. It is clear that the model, however useful it might be, will be subject to modification in detail as we learn more of the relevant enzymology, and perhaps, modification in principle as well, for it is, as yet, early days in the study of yeast cell cycle regulation.

ACKNOWLEDGMENTS

We thank Elizabeth Inhaber, Donna MacLeish, Gabrielle Préfontaine, and Isabelle Boisclair-Sarrazin for technical assistance. One of us (B. F. J.) thanks Dr. D. H. Williamson for stimulating discussion and initiation into the rites of *petite saccharomycenae*.

REFERENCES

1. Meissel, M. N., *Zentralbl. Bakteriol., Parasitenkd., Infektionskr. Hyg., Abt. 2* **88,** 449 (1933).
2. Pett, L. B., *Biochem. J.* **30,** 1438 (1936).
3. Stier, T. J. B., and Castor, J. G. B., *J. Gen. Physiol.* **25,** 229 (1941).
4. Whelton, R., and Phaff, H. J., *Science* **105,** 44 (1947).
5. Ephrussi, B., and Hottinguer, H. *Cold Spring Harbor Symp. Quant. Biol.* **16,** 75 (1951).
6. Slonimski, P. P., and Ephrussi, B., *Ann. Inst. Pasteur, Paris* **77,** 47 (1949).
7. Sherman, F.G.,, and Slonimski, P. P., *Biochim. Biophys. Acta* **90,** 1 (1964).
8. Yotsuyanagi, Y., *J. Ultrastruct. Res.* **7,** 141 (1962).
9. Ephrussi, B., Margerie-Hottinguer, H., and Roman, H., *Proc. Natl. Acad. Sci. U.S.A.* **41,** 1065 (1955).
10. Schatz, G., Halsbrunner, E., and Tuppy, H., *Biochem. Biophys. Res. Commun.* **15,** 127 (1964).
11. Tewari, K. K., Jayaraman, J., and Mahler, H. R., *Biochem. Biophys. Res. Commun.* **21,** 141 (1965).
12. Williamson, D. H., *Symp. Soc. Exp. Biol.* **24,** 247 (1970).
13. Nagley, P., and Linnane, A. W., *Biochem. Biophys. Res. Commun.* **39,** 989 (1970).

14. Mounolou, J. C., Jakob, H., and Slonimski, P. P., *Biochem. Biophys. Res. Commun.* **24,** 218 (1966).
15. Marcovich, H., *Ann. Inst. Pasteur, Paris* **81,** 452 (1951).
16. Moustacchi, E., and Marcovich, H., *C. R. Hebd. Seances Acad. Sci.* **256,** 5646 (1963).
17. Slonimski, P. P., *in* "Biochemical Aspects of the Biogenesis of Mitochondoria" (E. C. Slater *et al.,* eds.), p. 477. Adriatica Editrice, Bari, 1968.
18. Carnevalli, F., Morpurgo, G., and Tecce, G., *Science* **163,** 1331 (1969).
19. Rank, G. H., *Can. J. Genet. Cytol.* **12,** 129 (1970).
20. Mills, D. R., Peterson, R. L., and Spiegelman, S., *Proc. Natl. Acad. Sci. U.S.A.* **58,** 217 (1967).
21. Borst, P., and Kroon, A. M., *Int. Rev. Cytol.* **26,** 107 (1969).
22. Saunders, G. W., Gingold, E. B., Trembath, M. K., Lukins, H. B., and Linnane, A. W., *in* "Autonomy and Biogenesis of Mitchondria and Chloroplasts" (N. K. Boardman, A. W. Linnane, and R. M. Smillie, eds.), p. 185. North-Holland Publ., Amsterdam, 1971.
23. Preer, J. R., Jr., *Annu. Rev. Genet.* **5,** 361 (1971).
24. James, A. P., Johnson, B. F., Inhaber, E. R., and Gridgeman, N. T., *Mutat. Res.* **30,** 199 (1975).
25. Mortimer, R. K., and Johnston, J. R., *Nature* (*London*) **183,** 1751 (1959).
26. Nagai, S., Yanagishima, N., and Nagai, H., *Bacteriol. Rev.* **25,** 404 (1961).
27. Sherman, F. G., *J. Cell. Comp. Physiol.* **54,** 37 (1959).
28. Birky, C. W., Jr., *BioScience* **26,** 26 (1976).
29. Barton, A. A., *J. Gen. Microbiol.* **4,** 84 (1950).
30. Johnson, B. F., *Exp. Cell Res.* **39,** 577 (1965).
31. Johnson, B. F., and Gibson, E. J., *Exp. Cell Res.* **41,** 580 (1966).
32. Hayashibe, M., Sando, N., and Abe, N., *J. Gen. Appl. Microbiol.* **19,** 287 (1973).
33. Johnson, B. F., and Lu, C., *Exp. Cell Res.* **95,** 154 (1975).
34. Beran, K., Streiblová, E., and Lieblová J., *Proc. Int. Symp.* Yeasts, *2nd 1966* p. 353 (1969).
35. Campbell, A., *Bacteriol. Rev.* **21,** 263 (1957).
36. Williamson, D. H., and Fennell, D. J., *Mol. Gen. Genet.* **131,** 193 (1974).
37. Sena, E. P. Ph.D., Thesis, University of Wisconsin, Madison (1972).
38. Sena, E. P., Welch, J. W., Halvorson, H. O., and Fogel, S., *J. Bacteriol.* **123,** 497 (1975).
39. Wilkie, D., and Thomas, D. Y., *Genetics* **73,** 367 (1973).
40. Żuk, J., Zabarowska, D., Litwińska, J., Chlebowicz, E., and Biliński, T., *Acta Microbiol. Pol. Ser. A* **7,** 67 (1975).
41. Forster, J. L., and Kleese, R. A., *Mol. Gen. Genet.* **139,** 329 (1975).
42. Sebastian, J., Carter, B. L. A., and Halvorson, H. O., *J. Bacteriol.* **108,** 1045 (1971).
43. Dawes, I. W., and Carter, B. L. A., *Nature* (*London*) **250,** 709 (1974).
44. Negrotti, T., and Wilkie, D., *Biochim. Biophys. Acta* **153,** 341 (1968).
45. Rosin, M., Ph.D., Thesis, University of Toronto, Toronto (1976).
46. Mortimer, R. K., and Hawthorne, D. C., *in* "The Yeasts" (A. H. Rose and J. S. Harrison, eds.), Vol. 1, p. 448. Academic Press, New York, 1969.
47. Jinks, J. L., *in* "Methodology in Basic Genetics" (W. J. Burdette, ed.), p. 325. Holden-Day, San Francisco, California, 1963.
48. Petes, T. D., and Fangman, W. L., *Biochem. Biophys. Res. Commun.* **55,** 603 (1973).
49. Lee, E.-H., and Johnson, B. F., *J. Bacteriol* **129** 1066 (1977).
50. Sena, E. P., Welch, J.W., and Fogel, S., *Science* **194,** 433 (1976).
51. Mahler, H. R., Perlman, P. S., Feldman, F., and Bastos, R., *in* "Biomembranes: Architecture, Biogenesis, Biogenergetics, and Differentiation" (L. Packer, ed.), p. 3. Academic Press, New York, 1974.

12

Epidermal Proliferation in Lower Vertebrates

Charles W. Hoffman

I. INTRODUCTION

Continuous or periodic epidermal cellular proliferation occurs in all vertebrate classes (55). In fishes, there is a slow continual replacement of epidermal cells, which are sloughed separately or in small groups, whereas, in reptiles and amphibians, the outer layers are sloughed at intervals of a few days or weeks. This sloughing or molting process in vertebrates can be separated into three phases: (a) proliferation (usually of the basal layer), (b) differentiation, and (c) ecdysis or sloughing (shedding of the stratum corneum) (27). Perhaps because ecdysis is the most conspicuous of these phases, it has received the greatest attention and is often treated as a separate process. In contrast, relatively few studies have considered proliferation of the epidermal cells or the coordination of proliferation with the other phases or molting. Since it is essential that the loss of cells from the surface of the skin be balanced by proliferation of cells from below to maintain the thickness and integrity of the epidermis, the factors influencing the proliferative phase of molting are certainly worthy of investigation.

This chapter will review the regulation of epidermal cellular proliferation in three classes of vertebrates, fish, amphibians, and reptiles, with the greatest emphasis being placed on the hormonal regulation of epidermal proliferation.

II. FISH

In general, fish have a characteristic stratified vertebrate epidermis made up of cuticle-secreting cells, cuticularcytes, glandular goblet cells, and sensory cells. Although numerous studies have reported on the structure of the epidermis (9, 41–43, 52, 80), little is known about cyclic changes in the fish epidermis (78). Ecdysis may occur on a regular basis, and perhaps be under hormonal control (78), but extensive molting is reported only in the South African horsefish, *Agriopus* (34) and in the flying gunard, *Dactylopterus* by Heldt in 1927 (55).

Unlike the epidermis of other vertebrates, mitotic activity in fish epidermis is not restricted to the basal layer. In adult *Carassius*, [^{3}H]thymidine is incorporated throughout caudal fin and flank epidermis (five to eight layers), 90 minutes after injection of the label before cells could migrate from the basal layer (40).

Mitotic activity in fish epidermis may be geared to maintaining a layer of secretory cells at the skin surface and to replacing inactive dead cells (55). In this regard, hypophysectomy has been reported to decrease the number of mucous cells in many teleost (14, 68) and ectopic pituitary implants (68), releasing endogenous prolactin (5), or injection of prolactin (7) increases the number of mucous cells (for review, see 4). On the other hand, hypophysectomy and prolactin treatment failed to affect the number of mucous cells of *Tilapia mossambica* (6). In general, however, it would seem that prolactin is able to stimulate cell division in the teleost epidermis.

Hirano *et al.* (44) suggest that the stimulation of cell proliferation by prolactin may in fact reflect one of the primary actions of prolactin. This contention is supported by observations (see below) in both amphibians (45, 46a, 46b) and reptiles (20, 21, 62).

III. EPIDERMAL PROLIFERATION IN AMPHIBIANS

A. Epidermal Cell Cycle

Although many studies have considered mitotic activity in amphibian epidermis, few have dealt with the duration of the cell cycle or the effect on the cell cycle of factors that alter the molting process. The cell cycle in amphibians has most often been studied in regenerating urodele limbs (36, 86). Wallace and Maden (86) observed a cycle duration of 53 hours (S:38 hours G_1:10 hours, and G_2:5) in *Ambystoma mexicanum,* and Grillo (36) observed a cycle of 45 hours and 56 hours for newt limb blastema and connective tissue, respectively. The duration of the cell cycle in regenerating epidermis may differ from that of "normal" epidermis; however, these values are in agreement with cell cycle times for other amphibian tissues (36, 37, 89, 91).

To accurately determine the cell cycle using the method of Quastler and Sherman (72), it is critical that the label,[^{3}H]thymidine, is available for a defined period and that the population is asynchronous, i.e., does not exhibit a diurnal rhythm. These criterion may not be fulfilled in amphibian epidermis. Contradictory reports on the duration of the availability of[^{3}H]thymidine are present in the literature. Scheving and Chiakulas (75) reported that[^{3}H]thymidine is available for no more than 2 hours in larval *Ambystoma tigrinum,* whereas longer periods of availability (8 hours to 5 days) have been reported in *Notophthalamus viridescens* (38, 70) and *Am. mexicanum* (86). Wallace and Maden (86) recommend the use of a 1000 × chase of unlabeled thymidine 2 hours after the injection of label. In addition, since epidermal mitotic activity in amphibians is generally reported to be rhythmic (see below) it may be necessary to initiate the labeling at two or more time points (15) to precisely determine the parameters of the cell cycle.

The cell cycle of amphibian epidermal cells appears to be a neglected aspect of the molting process. As yet it is not known whether a single population of cells or more than one population of keratinocytes exists in the epidermis (18, 75). Furthermore, as will be demonstrated below, mitotic activity in the epidermis is influenced, directly or indirectly, by hormonal levels. It would be of interest to determine in what manner, e.g., shortening of the cycle duration or causing cells to enter or be removed from a G_0 or resting state, hormones alter the cell cycle.

B. Rhythmicity

Periodicity in epidermal mitotic activity has been reported in numerous mammals (for review, see 13), notably mice (12, 25), rats (8, 76), and man (26). Evidence for periodicity in the epidermal mitotic activity of amphibians has been reported for tadpole epidermis (66), larval *Xenopus laevis* tailfin epidermis (85) and larvae of *A. tigriniym* (17). In adult amphibians, periodicity has been reported in *Rana pipiens* (33) and *N. viridescens* (56). However, variability in the relation of the peak in mitotic activity to the light cycle is apparent. In *N. viridescens,* a single peak in mitotic activity was observed at the onset of the dark phase of the light cycle (2200–0200 hours, Fig. 1), whereas, in *R. pipiens,* the peak in mitotic activity occurred during the light phase of the cycle (33). In larval urodeles, a bimodal curve was observed with the greatest mitotic activity occurring during the light phase (17, 18, 75).

Periodicity was also noted in the incorporation of [^{3}H]thymidine in larval urodeles (17, 18, 75) and in adult *N. viridescens* (45) and *R. pipiens* (67). On the other hand, Jørgensen and Levi (49) were unable to find periodicity in the labeling index of the epidermis of the toad, *Bufo bufo.*

It is generally accepted that light serves as the dominant synchronizer of biologic rhythms and inversion of light cycle frequency causes a corresponding shift in rhythmic processes (for review, see 13, 39). However, in amphibians, although the light cycle appears to be a factor in the periodicity of epidermal mitotic activity the nature of its role is uncertain.

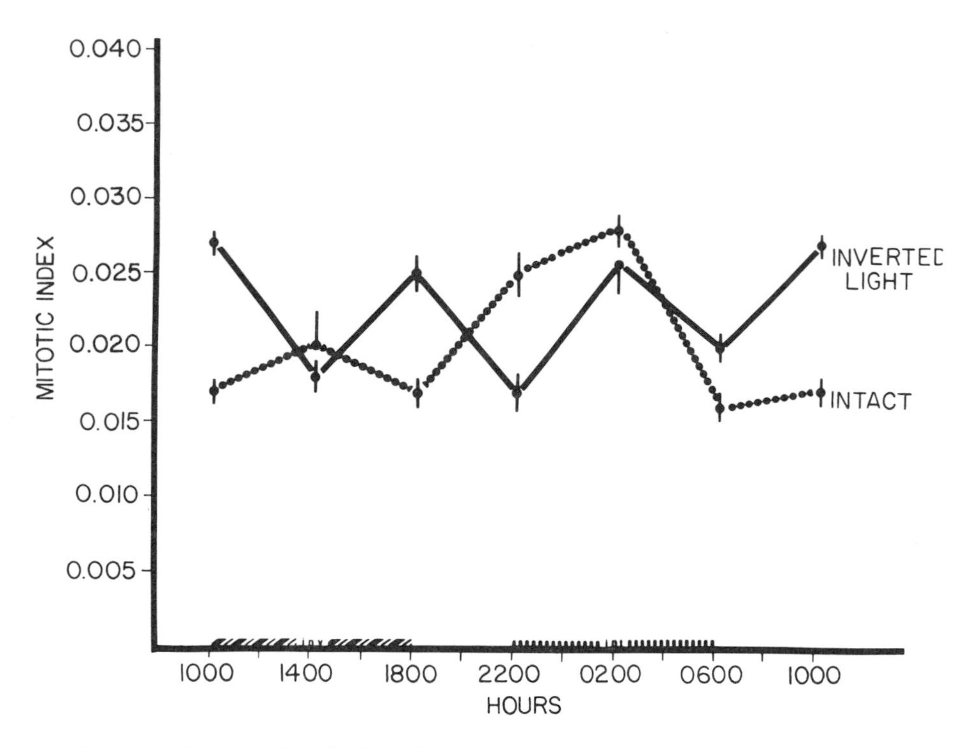

FIG. 1. Mitotic activity in the epidermis of intact newts on control and inverted light schedules. Vertical bars represent standard error of the mean; the horizontal bars of slanted lines and vertical lines on the time axis indicate the dark hours of the two light cycles. In the inverted light schedule, the dark hours were between 1000 and 1800 hours. For each point on control curve, $N = 18$; for inverted schedule, $N = 14$.

Parapinealectomy and blinding abolished the mitotic rhythm in *R. pipiens* (33), whereas, in tailfin epidermis of *Xenopus* larvae, the mitotic rhythm persisted in blinded tadpoles kept in a normal light cycle, but it was absent in tadpoles kept in continuous darkness suggesting an extraretinal pathway of regulation (85). In *R. pipiens* (33) and *N. viridescens* (46A) (Fig. 1), inversion of the light cycle failed to result in an inversion of the mitotic rhythm.

Few observations have been made on the relation of hormones to the mitotic rhythm in amphibians. In larval *T. torosa,* hypophysectomy reduced the magnitude of mitotic activity in the epidermis but did not abolish its rhythmicity (17). Conversely, in corneal epithelium of *R. pipiens,* hypophysectomy abolished the rhythm without producing a marked decrease in the incidence of mitosis (65). In *N. viridescens,* rhythmicity was not evident in the epidermis of newts 14 days after hypophysectomy, suggesting that hypophyseal hormones are necessary for mitotic rhythmicity (46A). An alternative explanation, in agreement with Chiakulas and Scheving (17) is that rhythmicity persists, but hypophysectomy lowers the magnitude of mitotic activity to such an extent that, in this case, by 14 days, the mitotic activity is so low that the rhythm is obscured, since a rhythm persists at 3 days but fades out at 7 days as the mitotic index declines.

Garcia-Arce and Mizell (33) found an inverse relation between mitotic activity and plasma corticoid levels (3) in *R. pipiens* and suggest that the antimitogenic effects of corticosteroids may be an important factor in controlling mitotic activity. In the urodeles, however, a correlation of plasma hormone levels and mitotic rhythmicity has not been reported.

C. Hormonal Effects on Epidermal Mitosis

The role of hormones in the sloughing phase of molting is well documented for both anurans and urodeles (53). Hormonal effects on the proliferative phase, however, have received less attention, and some results have been conflicting. Hypophysectomy diminished epidermal proliferation in amputated limbs of *T. phyrhogaster* (47) and in the epidermis of urodele larvae (17). On the other hand, Vellano *et al.* (81) found no difference in the epidermal mitotic indices of intact and hypophysectomized crested newts, and Jørgensen *et al.* (50) have suggested that the accumulation of unshed sloughs in hypophysectomized amphibians indicates that the pituitary is not necessary for epidermal proliferation. Jørgensen *et al.* (50) further suggest that hormones merely exert a "permissive" action, allowing the rhythmic processes inherent in the skin to be expressed. Hoffman and Dent (46b) have studied the effects of pituitary hormones on the proliferative phase of molting in the red-spotted newt, *N. viridescens*. These results will in part be reviewed here.

In the newt, hypophysectomy results in a decline in the mean mitotic index (the mean of mitotic indices determined during a 24-hour period) (Fig. 2). The decline in the mean mitotic index of the epidermis is accompanied by reductions in the mean number of cell layers and the epidermal thickness (Fig. 2), indicating that although slough formation continues after the removal of the pituitary gland, replacement of cells does not keep pace with keratinization. The decrease in the mitotic index that follows hypophy-

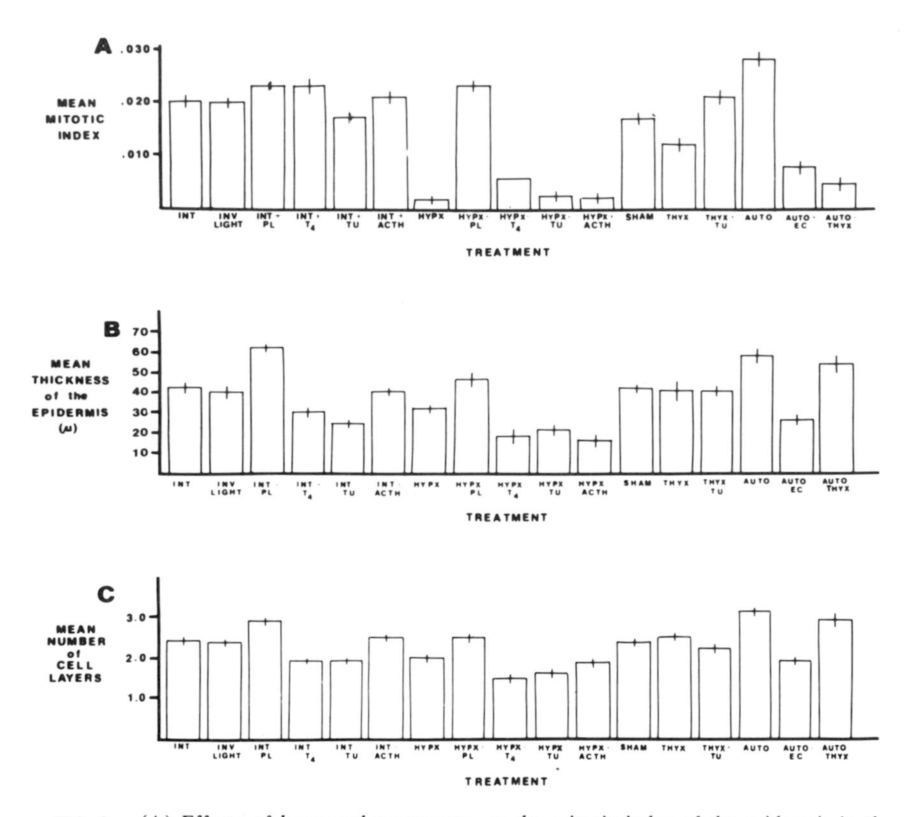

FIG. 2. (A) Effects of hormonal treatments on the mitotic index of the epidermis in the red-spotted newt. Vertical line through center of bar represents the standard error of the mean. Int, intact control; Inv, inverted light cycle; P_1, prolactin; T_4 = thyroid hormone; TU, thiourea; EC, ergocornine; Hypx, hypophysectomy (21 days); Thyx, thyroidectomy. (B) Effects of hormonal treatment on the thickness of the epidermis, the layer of living cells between the basal lamina and the stratum corneum was used as the index of epidermal thickness. (C) Effects of hormonal treatments on the number of cell layers in the epidermis. See Table I for numbers of animals in (A), (B), and (C).

sectomy can not be attributed to the stress of the operation or to the fact that food is refused after 7–10 days, since the mean mitotic index was significantly greater in sham-operated newts that had been starved for 5 weeks (Fig. 2) than in hypophysectomized newts. It would appear that in the newt, some hypophyseal principle is required for epidermal proliferation.

It has been suggested by Tassava (79) that prolactin stimulates cell division in the epidermis of the newt. The preponderance of evidence supports this view, although conflicting reports are found (81). Increased mitotic rates have been reported in the skins of prolactin-treated newts (16, 35) and in hypophysectomized urodele larvae with ectopically implanted pituitary glands (17). In the newt, injection of prolactin in intact and hypophysectomized newts and autografting of the pituitary, which produces elevated levels of prolactin (28, 63), increases the mean mitotic index and increases the number of cell layers and the thickness of the epidermis (Fig. 2).

The drug, ergocornine, has been shown to inhibit specifically the secretion of prolactin in urodeles (71) and in mammals (64, 90) without altering thyroid activity. Intraperitoneal injection of ergocornine into autografted newts, in contradistinction to the results obtained by autografting and by administration of prolactin, decreased the mean mitotic index, the number of cell layers, and the thickness of the epidermis.

It is not certain whether prolactin acts directly on the epidermis to stimulate mitosis or indirectly by an increase in general metabolism. The latter indirect action seems more probable. Prolactin has been shown to enhance food consumption (74), to increase body weight (73, 87), and to have a growth-promoting effect in the red-spotted newt (10) and in amphibian tadpoles (32). Tassava (79) has also demonstrated that regeneration is more rapid in well-fed than in starved newts.

It is the available level of thyroid hormone that regulates the rate at which sloughs are shed in urodeles (1, 2, 24, 28, 48, 69). The action of the thyroid on the proliferative phase of molting in urodeles, however, is more difficult to assess. That thyroid hormone stimulates proliferation is suggested by an increase in the mean mitotic index in intact newts treated with thyroid hormone and, conversely, by a decrease in the mean mitotic index in thyroidectomized newts and in autografted newts that had also been thyroidectomized (Fig. 2). However, thyroid hormone alone did not restore mitotic activity in hypophysectomized newts, indicating that some hypophyseal factor, probably prolactin, must be present if thyroxine is to produce a stimulatory effect on epidermal proliferation. That factor must continue to stimulate proliferation in the absence of the thyroid, since the decrease in mitotic activity produced by removal of the thyroid is not nearly as great as that produced by hypophysectomy (Fig. 2).

It is probable that thyroid hormone affects proliferation indirectly by increasing or decreasing the rate at which sloughs are shed and that a regulative mechanism inherent within the epidermis acts to coordinate the sloughing and proliferative phases. Although tentative, this view is supported by several observations. The mean mitotic index declined in thyroidectomized newts, when unshed sloughs accumulated, even though the pituitary gland presumably continued to produce prolactin. When sloughing is reinstituted, however, by treating thyroidectomized newts with thiourea, the mean mitotic index rose to a level equivalent to that of intact control newts (Fig. 2). Thiourea is usually employed to inhibit thyroid activity. In the newt, however, it also has the peculiar property of inducing sloughing thereby allowing sloughs to be removed without restoring thyroid activity. However, thiourea, like thyroxine, failed to restore mitotic activity in hypophysectomized newts (Fig. 2), indicating that the stimulatory effect of sloughing on proliferation is an accessory one and that some hypophyseal factor must be present for proliferation to proceed.

The increase in sloughing and in mitotic activity caused by thyroid hormone in intact newts is accompanied by a decrease in the number of cell layers and in epidermal thickness. Apparently, the rate of sloughing, in response to high levels of exogenous thyroxine, becomes excessive, such that it can no longer be matched by increases in proliferation, and cells are removed from the transition zone more rapidly than they are added to it.

These results are not directly in agreement with the view of Jørgensen *et al.* (50) that hormones simply exert a "permissive" action on the molting process. Prolactin appears to be able to stimulate epidermal proliferation in the newt, resulting in increase in the thickness of the epidermis, whereas high levels of thyroid hormone result in a thinning of the epidermis. Such changes in epidermal thickness may be of significance in osmoregulation (11).

The possible involvement of the pituitary–adrenal axis in the regulation of epidermal proliferation should also be considered. As previously stated, and inverse relationship between mitotic rates and levels of plasma corticoids has been found in the epidermis of *R. pipiens* (3, 33). Although a direct involvement of the pituitary–adrenal axis in limb regeneration in newts (77) has been questioned (79, 88). Vethamany-Globus and Liversage (82–84) have stressed the stimulatory effect of adrenal corticoids on the islets of the pancreas and have presented evidence to show that insulin promotes regeneration of the forelimb and tail in the newt. Also, a phasing effect of corticosterone on prolactin-induced water drive in efts has been demonstrated (51). In preliminary studies, however, ACTH did not have an effect on the mitotic activity in the epidermis of intact or hypophysectomized red-spotted newts (46B).

IV. REPTILES

The epidermis and its renewal in reptiles has been studied most extensively in the squamates (lizards and snakes). In the epidermal sloughing cycle of squamates, a keratinized structure called an "epidermal generation" is periodically shed from the entire body (54, 58, 59, 61, 62). This sloughing cycle can be divided into two major parts, a resting phase and a renewal phase, and a minor shedding phase of shorter duration (1–12 hours) (62). These phases can in turn be subdivided. The resting phase includes a "perfect resting condition," during which little if any cell production occurs, and the renewal phase can be separated into five stages covering the differentiation of the inner generation (54, 58–61).

During the sloughing cycle of the lizard, peaks in mitotic activity may occur: (a) as the outer epidermal generation is completed, (b) around the time of shedding, and (c) during the renewal phase as the presumptive mesopopulation is laid down (62). Although ^{3}H-thymidine has been used to follow the fate of the cell in the basal layer (29, 30) as yet, no studies have considered parameters of the cell cycle during the peaks in mitotic activity.

The principal effect of hormones on the molting cycle in reptiles appears to be on the duration of the resting phase. In snakes, removal of the thyroid gland increases the rate of sloughing by reducing the duration of the resting phase, and conversely, treatment with thyroid hormone increases the duration of the resting phase, thereby decreasing the rate of sloughing (22, 57, 62). The opposite situation exists in lizards where thyroidectomy and hypophysectomy increase the length of the resting phase, decreasing the rate of sloughing. Treatment of intact, thyroidectomized, and hypophysectomized lizards with thyroxine, in turn, increases the rate of sloughing by decreasing the duration of the resting phase (19, 62). In addition to thyroid hormone, prolactin has been shown to act alone (20, 21) and synergistically with thyroid hormone (23), directly on the skin to shorten the duration of the resting phase in the lizard (21). Exogenous prolactin, however, has not been found to have an effect on sloughing in snakes (22). *In vitro* studies with lizard skin also indicate that hormones affect the resting phase of molting and, thereby, initiate or inhibit proliferation, but that proliferation itself and the pattern of differentiation during the renewal phase are not subject to hormonal control (31, 62).

In the lizard, proliferation and the pattern of differentiation in which alternate layers of α- and β-keratins are synthesized appears to be controlled by an intraepidermal feedback between the outer and inner epidermal layers (29). This feedback may act at the level of the germinal layers. When basal cells are labeled with ^{3}H-thymidine just prior to the emergence of lacunar cells, they remain in the basal layer for several days, although five to seven layers of nonlabel cells, synthesizing β-keratin are added above the basal layer (29, 30). The emergence of labeled cells coincides with the emergence of cells that will form α-keratin, suggesting that the length of time a cell spends in the basal layer and the number of times it divides may determine the pattern of differentiation (29). The mechanism by which the feedback functions, however, remains uncertain.

V. CONCLUDING REMARKS

The status of the understanding of regulation of epidermal proliferation in the lower vertebrate is far behind the level attained in the mammalian epidermis. It is fair to state that, in the lower vertebrates, particularly the fish and amphibians, because of conflicting data in the literature, a basic understanding of the factors affecting epidermal proliferation within the molting process has not yet been attained. Fortunately, in the reptiles a more coherent picture of the molting process has emerged, although many aspects here also remain undertain. It is however, essential that the fundamental aspects of the proliferative phase of the molting cycle be understood before mechanistic approaches be attempted.

It appears certain that the molting cycle in reptiles and amphibians is under neurohypophyseal influence, which, in turn, is affected by environmental factors. In squamates, this influence is primarily at the proliferative phase and sloughing phase are subject to hormonal influence.

REFERENCES

1. Adams, A. E., and Richards, L., *Anat. Rec.* **44,** 222 (1929).
2. Adams, A. E., Kuder, A., and Richards, L., *J. Exp. Zool.* **63,** 1 (1932).
3. Akin, D. P., and Mizell, S., *Fed. Proc., Fed. Am. Soc. Exp. Biol.* **30,** 610 (1971).
4. Ball, J. N., *Fish Physiol.* **2,** 207 (1969).
5. Ball, J. N., Olivereau, M., Slicher, A. M., and Kallman, K. D., *Philos. Trans. R. Soc. London, Ser. B* **249,** 69 (1965).
6. Bern, H. A., *Science* **158,** 455 (1967).
7. Blum, V., and Fiedler, K., *Gen. Comp. Endocrinol.* **5,** 186 (1965).
8. Blumenfeld, C. M., *Science* **131** 1039 (1939).
9. Brown, G. A., and Welling, S. R., *Z. Zellforsch. Mikrusk. Anat.* **103,** 149 (1970).
10. Brown, P. S., and Brown, S. C., *J. Exp. Zool.* **178,** 29 (1971).
11. Brown, P. S., and Brown, S. C., *Gen. Comp. Endocrinol.* **20,** 456 (1973).
12. Bullough, W. S., *Proc. R. Soc. London, Ser. B* **135,** 212 (1945).
13. Bunning, E., "The Physiological Clock." Springer-Verlag, Berlin and New York, 1967.
14. Burden, C. E., *Biol. Bull.* **110,** 8 (1965).
15. Burns, E. R., Scheving, L. E., Fawcett, D. F., Gibbs, W. M., and Galatzan, R. E., *Anat. Rec.* **184,** 265 (1975).

16. Chadwick, C. S., and Jackson, H. R., *Anat. Rec.* **101,** 78 (1948).
17. Chiakulas, J. J., and Scheving, L. E., *In* "The Cellular Aspects of Biorhythms" (H. von Mayersbach, ed.) p. 155. Springer-Verlag, Berlin and New York, 1965.
18. Chiakulas, J. J., and Scheving, L. E., *Exp. Cell Res.* **44,** 256 (1966).
19. Chiu, K. W., and Phillips, J. G., *J. Endocrinol.* **49,** 611 (1971).
20. Chiu, K. W., and Phillips, J. G., *J. Endocrinol.* **49,** 619 (1971).
21. Chiu, K. W., and Phillips, J. G., *J. Endocrinol.* **49,** 625 (1971).
22. Chiu, K. W., and Phillips, J. G., *Copeia No. 1,* p. 158 (1972).
23. Chiu, K. W., and Phillips, J. G., *Gen. Comp. Endocrinol.* **19** 592 (1972).
24. Clark, N. B., and Kaltenbach, J. C., *Gen. Comp. Endocrinol.* **1,** 513 (1961).
25. Cooper, Z. K., and Franklin, N., *Anat. Rec.* **78,** 1 (1940).
26. Cooper, Z. K., and Schift, A., *Proc. Soc. Exp. Biol. Med.* **39,** 323 (1958).
27. Dent, J. N., *Am. Zool.* **15,** 923 (1975).
28. Dent, J. N., Eng, L. A., and Forbes, M. S., *J. Exp. Zool.* **184,** 369 (1973).
29. Flaxman, B. A., *Am. Zool.* **12,** 13 (1972).
30. Flaxman, B. A., and Maderson, P., *J. Exp. Zool.* **183,** 209 (1973).
31. Flaxman, B. A., Maderson, P., Szabo, G., and Roth, S. I., *Dev. Biol.* **18,** 354 (1968).
32. Frye, B. E., Brown, P. S., and Snyder, B. W., *Gen. Comp. Endocrinol., Suppl.* **3,** 209 (1972).
33. Garcia-Arce, H., and Mizell, S., *Comp. Biochem. Physiol. A* **42,** 501 (1972).
34. Gilchrist, J. D. F., *J. Microsc. Sci.* [N.S.] **64,** 575 (1920).
35. Grant, W. C., and Grant, J. A., *Biol. Bull.* **114,** 1 (1958).
36. Grillo, R. S., *Oncology* **25,** 347 (1971).
37. Grillo, R. S., and Urso, P., *Oncology* **22,** 208 (1968).
38. Grillo, R. S., and Urso, P., and O'Brian, D. M., *Exp. Cell Res.* **37** 683 (1965).
39. Halberg, F., "Advances in Experimental Medicine and Biology, Biological Rhythms and Endocrine Function." (L. W. Hedlund, J. M. Franz, and A. D. Kenny, eds.), p. 1. Plenum, New York, 1975.
40. Henrikson, R. C., *Experientia* **23,** 357 (1967).
41. Henrikson, R. C., and Matoltsy, A. G., *J. Ultrastruct. Res.* **21,** 194 (1968).
42. Henrikson, R. C., and Matoltsy, A. G., *J. Ultrastruct. Res.* **21,** 213 (1968).
43. Henrikson, R. C., and Matoltsy, A. G., *J. Ultrastruct. Res.* **21,** 222 (1968).
44. Hirano, T., Hayashi, S., and Utida, S., *J. Endocrinol* **56,** 591 (1973).
45. Hoffman, C. W., and Dent, J. N. *Am. Zool.* **13,** 1281 (1973).
46a. Hoffman, C. W. and Dent, J. N. (1977). *Gen. Comp. Endocrinol.* **32,** 000 (1977).
46b. Hoffman, C. W. and Dent, J. N. (1977). *Gen. Comp. Endocrinol.* **32:** 000 (1977).
47. Inoué S., *Endocrinol. Jpn.* **3,** 158 (1956).
48. Jørgensen, C. B., and Larsen, L. O., *Nature (London)* **185,** 244 (1960).
49. Jørgensen, C. B., and Levi, H., *Comp. Biochem. Physiol. A* **52,** 55 (1975).
50. Jørgensen, C. B., Larsen, L. O., and Rosenkilde, P., *Gen. Comp. Endocrinol.* **5,** 248 (1965).
51. Joseph, M. M., *In* "Chronobiology (L. E. Scheving, F. Halberg, and J. E. Pauly, (eds.), p. 197–201. Igaku Shoin Ltd., Tokoyo, 1974.
52. Lanzing, W. J., and Wright, R. G., *Cell Tissue Res.* **154,** 251 (1974).
53. Larsen, L. O., *In* "Physiology of the Amphibia" (B. Lofts, ed.), Vol. 3, p. 000. Academic Press, New York, 1976.
54. Lillywhite, H. B., and Maderson, P., *J. Morph.* **124,** 1 (1968).
55. Ling, J. K., (1972). *Am. Zool.* **12,** 77 (1972).
56. Litwiller, R., *Growth* **4,** 168 (1940).
57. Lynn, W. G., *In* "Biology of the Repilia" (C. Gans and T. S. Parsons, eds.), Vol. 3, p. 201. New York, 1970.

58. Maderson, P., *J. Zool.* **146,** 98 (1965).
59. Maderson, P., *J. Morphol.* **119,** 39 (1966).
60. Maderson, P., *Copeia* p. 743 (1967).
61. Maderson, P., and Licht, P., *J. Morph.* **123,** 157 (1967).
62. Maderson, P., Chiu, K. W., and Phillips, J. G., *Mem. Soc. Endocrinol.* **28,** 259 (1970).
63. Masur, S. K., (1962). *Am. Zool.* **2,** 538 (1962).
64. Meites, J., Lu, K. H., Wuttke, W., Welsch, C. W., Nagasawa, H., and Quadri, S. K., *Recent Prog. Horm. Res.* **28,** 471 (1972).
65. Mizell, S., and Hutto, A., *Comp. Biochem. Physiol. A* **39,** 227 (1971).
66. Mollerberg, H., *Acta Anat.* **4,** (1947).
67. Morgan, W. W., and Mizell, S., *Comp. Biochem. Physiol. A* **38** 591 (1971).
68. Ogawa, M., and Johansen, P. H., *Can. J. Zool.* **45,** 885 (1967).
69. Osborn, C. M., *Anat. Rec.* **66,** 257 (1936).
70. O'Steen, W. K., and Walker, B. E., *Anat. Rec.* **139,** 547 (1961).
71. Platt, J. E., *Gen. Comp. Endocrinol.* **28,** 71 (1976).
72. Quastler, H., and Sherman, F. G., *Exp. Cell Res.* **17,** 420 (1959).
73. Schauble, M. K., and Nentwig, M. R., *J. Exp. Zool.* **187,** 335 (1974).
74. Schauble, M. K., and Tyler, D. B., (1972). *J. Exp. Zool.* **182,** 41(1972).
75. Scheving, L. E., and Chiakulas, J. J., *Exp. Cell Res.* **39,** 161 (1965).
76. Scheving, L. E., and Pauly, J. E., (1960). *Acta Anat.* **43,** 337 (1960).
77. Schotté, O. E., *In* "Molecular and Cellular Structures (D. Rudnick, ed.), p. 161. Ronald Press, New York, 1961.
78. Spearman, R., "*The Integument.*" Cambridge Univ. Press, London and New York, 1973.
79. Tassava, R. A., *J. Exp. Zool.* **170,** 33 (1969).
80. Van Oosten, J. G., *In* "The Physiology of Fishes" (M. E. Brown, ed.), Academic Press, Vol. 1, p. 207. New York, 1957.
81. Vellano, C., Lodi, G., Bani, G., Sacerdote, M., and Mazzi, V., *Monit. Zool. Ital.* (N.S.) **4,** 115 (1970).
82. Vethamany-Globus, S., and Liversage, R. A., *J. Embryol. Exp. Morphol.* **30,** 397 (1973).
83. Vethamany-Globus, S., and Liversage, R. A., *J. Embryol. Exp. Morphol.* **30,** 415 (1973).
84. Vethamany-Globus, S., and Liversage, R. A., *J. Embryol. Exp. Morphol.* **30,** 427 (1973).
85. Wakahara, M., *Neuroendocrinology* **9,** 267 (1972).
86. Wallace, H., and Maden, M., *J. Cell Sci.* **20,** 539 (1976).
87. Waterman, A. J., *Am. Zool.* **5,** 237 (1965).
88. Wilkerson, J. A., *J. Exp. Zool.* **154,** 223 (1963).
89. Wilson, B. G. (1975). *Chromosoma* **51,** 213 (1975).
90. Wuttke, W., Cassell, E., and Meites, J., *Endocrinology* **88,** 734 (1971).
91. Zalik, S. E., and Yamada, T., *J. Exp. Zool.* **165,** 385 (1967).

Author Index

Numbers in parentheses are reference numbers and indicate that an author's work is referred to although his name is not cited in the text. Numbers in italics show the page on which the complete reference is listed.

H

M

N

O

S

T

U

V

Subject Index

CELL BIOLOGY: A Series of Monographs

EDITORS

G. M. Padilla, G. L. Whitson, and I. L. Cameron (editors). THE CELL CYCLE: Gene-Enzyme Interactions, 1969

A. M. Zimmerman (editor). HIGH PRESSURE EFFECTS ON CELLULAR PROCESSES, 1970

I. L. Cameron and J. D. Thrasher (editors). CELLULAR AND MOLECULAR RENEWAL IN THE MAMMALIAN BODY, 1971

I. L. Cameron, G. M. Padilla, and A. M. Zimmerman (editors). DEVELOPMENTAL ASPECTS OF THE CELL CYCLE, 1971

P. F. Smith. The BIOLOGY OF MYCOPLASMAS, 1971

Gary L. Whitson (editor). CONCEPTS IN RADIATION CELL BIOLOGY, 1972

Donald L. Hill. THE BIOCHEMISTRY AND PHYSIOLOGY OF *TETRAHYMENA*, 1972

Kwang W. Jeon (editor). THE BIOLOGY OF AMOEBA, 1973

Dean F. Martin and George M. Padilla (editors). MARINE PHARMACOGNOSY: Action of Marine Biotoxins at the Cellular Level, 1973

Joseph A. Erwin (editor). LIPIDS AND BIOMEMBRANES OF EUKARYOTIC MICROORGANISMS, 1973

A. M. Zimmerman, G. M. Padilla, and I. L. Cameron (editors). DRUGS AND THE CELL CYCLE, 1973

Stuart Coward (editor). DEVELOPMENTAL REGULATION: Aspects of Cell Differentiation, 1973

I. L. Cameron and J. R. Jeter, Jr. (editors). ACIDIC PROTEINS OF THE NUCLEUS, 1974

Govindjee (editor). BIOENERGETICS OF PHOTOSYNTHESIS, 1975

James R. Jeter, Jr., Ivan L. Cameron, George M. Padilla, and Arthur M. Zimmerman (editors). CELL CYCLE REGULATION, 1978

A 8
B 9
C 0
D 1
E 2
F 3
G 4
H 5
I 6
J 7